SpringerBriefs in Computer Science

For further volumes:
http://www.springer.com/series/10028

SpringerBriefs in Computer Science

Series Editors

Stan Zdonik
Peng Ning
Shashi Shekhar
Jonathan Katz
Xindong Wu
Lakhmi C. Jain
David Padua
Xuemin Shen
Borko Furht
V.S. Subrahmanian
Martial Hebert
Katsushi Ikeuchi
Bruno Siciliano

For further volumes:
http://www.springer.com/series/10028

Sherin Abdel Hamid · Hossam S. Hassanein
Glen Takahara

Routing for Wireless
Multi-Hop Networks

Springer

Sherin Abdel Hamid
Telecommunications Research Lab
Queen's University
Kingston, ON
Canada

Glen Takahara
Telecommunications Research Lab
Queen's University
Kingston, ON
Canada

Hossam S. Hassanein
Telecommunications Research Lab
Queen's University
Kingston, ON
Canada

ISSN 2191-5768   ISSN 2191-5776 (electronic)
ISBN 978-1-4614-6356-6   ISBN 978-1-4614-6357-3 (eBook)
DOI 10.1007/978-1-4614-6357-3
Springer New York Heidelberg Dordrecht London

Library of Congress Control Number: 2012955038

Printed on acid-free paper

Springer is part of Springer Science+Business Media (www.springer.com)

Preface

Wireless communication provides great advantages that are not available through their wired counterparts such as flexibility, ease of deployment and use, cost reductions, and convenience. However, these advantages come at the expense of some drawbacks, the most prominent of which is the limitation of the transmission range of wireless nodes. This limitation is due to the characteristics inherent in wireless communication such as fading, frequency reuse, noise, interference, and receiver sensitivity. As a result, a wireless node can only communicate directly with nodes within its transmission range. In order to communicate with out-of-range nodes when wireless nodes are deployed in an ad hoc setup with no infrastructure, a wireless node has to depend on other intermediate nodes for relaying its messages until they reach the intended destination. This communication paradigm is known as "multi-hop" communication, where each node can act as a source, a destination, or a router relaying messages.

In a wireless multi-hop network, one of the important challenges is how to route packets efficiently. The availability of many intermediate nodes between a source and a destination results in having many optional paths/routes to follow. The challenge is to pick the optimal path that satisfies the needed performance requirements, and this is the responsibility of a routing protocol. Choosing an optimal path from a source to a destination can be done by optimizing one or more routing metrics (such as number of hops, distance, delay, packet loss rate, and energy consumption). The selection metric is chosen based on application requirements such as delay-sensitivity or on constraints such as limited energy or frequent topology changes.

There are four wireless network paradigms falling under the category of wireless multi-hop networks. These paradigms are Mobile Ad Hoc Networks (MANETs), Wireless Sensor Networks (WSNs), Wireless Mesh Networks (WMNs), and Vehicular Ad-Hoc Networks (VANETs). In these four network paradigms, routing plays a vital and critical role and is considered one of the most important design elements of these networks.

Following a component-based approach, routing protocols for wireless multi-hop networks can be decomposed into smaller functional components. A routing protocol can be a combination of some or all of these components depending on the characteristics of the network that this protocol is proposed for and on the application requirements as well. Some of these routing components are core and should be a part of the skeleton of any routing protocol. These fundamental components are route discovery, route selection, and route representation and data forwarding. Some other components are network-dependent and will be activated and used only based on network needs. Examples of such auxiliary components are route maintenance and route energy efficiency.

Being categorized as wireless multi-hop networks, the four aforementioned network paradigms share some commonalities in terms of their routing function. However, as each of these network paradigms has its own unique characteristics and environment/application needs, each has some distinct aspects that distinguish its routing approaches from the others. The target of this brief is to show the unifications and distinctions of the routing functions of the various multi-hop network paradigms.

Over the past years, many surveys have addressed routing protocols for each of the aforementioned wireless multi-hop networks. Yet, there are many questions that need to be answered: Why is there not a unified set of routing protocols that can be used for all these types of networks based on the fact that they are all wireless multi-hop networks? Why does each type of network require the design of its own routing protocols? What aspects distinguish each of these networks in terms of routing? etc. In addition to discussing the commonalities, this brief answers these questions with the objective of showing the distinguishing features of the routing functions of the various wireless multi-hop networks.

The brief is organized as follows: as a common ground, in Chap. 1, we present an overview of wireless multi-hop networks along with a brief introduction to each of the four aforementioned wireless multi-hop network paradigms. In Chap. 2, we show the *"unifying features"* of routing by presenting an overview of routing in wireless multi-hop networks, its basic concepts, and the various routing components that can form a wireless multi-hop routing protocol. Both core and auxiliary components are highlighted. In addition, we introduce a generic routing model that can be the foundation of the wireless multi-hop routing function and can be inherited by any wireless multi-hop routing protocol. In Chap. 3, to highlight the *"distinguishing features"*, we present the requirements and design considerations of each of the four aforementioned wireless multi-hop network paradigms. Also, the popular classification of routing protocols for each network paradigm is presented. Furthermore, we discuss the routing components that should be activated and included as core parts of a routing protocol for each network paradigm along with some various functionalities of each component and some examples of routing protocols that adopt these functionalities. In addition, we summarize the distinctions part by providing an abstraction for the general routing functionalities of each of the four network paradigms. Finally, in Chap. 4, we present some concluding remarks along with some potential open issues.

This brief is intended for readers interested in getting an overview about this field of research and for researchers interested in further research and contributions. It provides an exhaustive view of the wireless multi-hop routing components and aspects along with in-depth discussions about the wireless multi-hop network paradigms in terms of the commonalities and distinctions of their routing functions. We hope that this brief will be an inspiration for many ideas and contributions in the near future and will open doors for fruitful research avenues.

Kingston, Ontario, Canada, August 2012   Sherin Abdel Hamid
                                                                     Hossam S. Hassanein
                                                                        Glen Takahara

Acknowledgments

This research is funded by a grant from the Ontario Ministry of Economic Development and Innovation under the Ontario Research Fund-Research Excellence (ORF-RE) program.

Contents

Abbreviations

ADAS	Advanced Driver Assistance Services
ADV	Advertisement
AODV	Ad-hoc On-Demand Distance Vector
AODV-ST	Ad-hoc On-Demand Distance Vector-Spanning Tree
ARPANET	Advanced Research Projects Agency Network
A-STAR	Anchor-based Street and Traffic Aware Routing
BLC	Bottleneck Link Capacity
BS	Base Station
CAR	Capacity-Aware Routing
CAR	Connectivity-Aware Routing
CBF	Contention-Based Forwarding
CBRP	Cluster Based Routing Protocol
CEDAR	Core-Extraction Distributed Ad Hoc Routing
CFR	Congestion Free Routing
CGSR	Cluster-head Gateway Switch Routing
CH	Cluster-Head
DAG	Directional Acyclical Graph
DARPA	Defense Advanced Research Projects Agency
DCAR	Distributed Coding-Aware Routing
DD	Directed Diffusion
DREAM	Distance Routing Effect Algorithm for Mobility
DSDV	Destination Sequence Distance Vector
DSR	Dynamic Source Routing
DV	Distance Vector
ENT	Effective Number of Transmissions
ETT	Expected Transmission Time
ETX	Expected Transmission Count
ExOR	Extremely Opportunistic Routing
FSR	Fisheye State Routing
GA	Genetic Algorithm
GBR	Gradient-Based Routing

GEAR	Geographical and Energy Aware Rrouting
GPCR	Greedy Perimeter Coordinator Routing
GPS	Global Positioning System
GPSR	Greedy Perimeter Stateless Routing
GPSR+AGF	Greedy Perimeter Stateless Routing + Advanced Greedy Forwarding
GRANT	Greedy Routing with Abstract Neighbor Table
GSR	Geographic Source Routing
HWMP	Hybrid Wireless Mesh Protocol
IARP	IntrA-zone Routing Protocol
IERP	IntEr-zone Routing Protocol
IRMA	Integrated Routing and MAC Scheduling Algorithm
ITS	Intelligent Transportation System
LAN	Local Area Network
LAR	Location Aided Routing
LBAR	Load-Balanced Ad-hoc Routing
LCMP	Light Client Management routing Protocol
LEACH	Low Energy Adaptive Clustering Hierarchy
LIBRA	Load and Interference Balanced Routing Algorithm
LQSR	Link Quality Source Routing
LS	Link State
MAC	Medium Access Control
MANET	Mobile Ad-hoc Network
MCFA	Minimum Cost Forwarding Algorithm
MIC	Metric of Interference and Channel-switching
MMESH	Multipath Mesh
MORE	MAC-independent Opportunistic Routing & Encoding
MPR	Multi-Point Relay
MR-LQSR	MultiRadio-Link Quality Source Routing
NCLBR	Node Centric Load Balancing Routing
OBU	On-Board Unit
OLSR	Optimized Link State Routing
ORR	Orthogonal Rendezvous Routing
OTR	Optimized Tree-based Routing
PEGASIS	Power-Efficient Gathering in Sensor Information Systems
PRENET	Packet Radio Networks
QoS	Quality of Service
REAR	Reliable Energy Aware Routing
ROMER	Resilient Opportunistic Mesh Routing
RREP	Route Reply
RREQ	Route Request
RSU	Road Side Unit
SADV	Static-Node Assisted Adaptive
SPIN	Sensor Protocol for Information via Negotiation
TBR	Tree-Based Routing
TORA	Temporally Ordered Routing Algorithm

VADD	Vehicle-Assisted Data Delivery
VANET	Vehicular Ad-Hoc Networks
V2I	Vehicle-to-Infrastructure
V2V	Vehicle-to-Vehicle
WAVE	Wireless Access for Vehicular Environment
WCETT	Weighted Cumulative Expected Transmission Time
WMN	Wireless Mesh Network
WMR	Wireless Mesh Router
WMSN	Wireless Multimedia Sensor Network
WRP	Wireless Routing Protocol
WSN	Wireless Sensor Network
ZHLS	Zone-based Hierarchical Link State
ZRP	Zone Routing Protocol

Chapter 1
Introduction to Wireless Multi-Hop Networks

Abstract Although wireless communication has brought many benefits since its introduction, it has some drawbacks compared to its wired counterpart. Wireless communication suffers from interference, low bandwidth availability, low data rates compared to wire line, and signal fading. Such drawbacks led to limitations in the transmission range of wireless devices. For a node to send a packet to a destination out of its transmission range, the node should depend on some intermediate node(s) for relaying the packet. Such a paradigm is known as multi-hop communication and the wireless networks adopting this communication paradigm are known as wireless multi-hop networks. Four network paradigms can be classified as wireless multi-hop networks. These paradigms are: Mobile Ad-Hoc Networks (MANETs), Wireless Sensor Networks (WSNs), Wireless Mesh Networks (WMNs), and Vehicular Ad-Hoc Networks (VANETs). In this chapter, we will present an overview of wireless multi-hop networks along with brief introductions to these four wireless multi-hop network paradigms.

Keywords Wireless multi-hop networks · Mobile ad-hoc networks · Wireless sensor networks · Wireless mesh networks · Vehicular ad-hoc networks

1.1 Overview

Generally, wireless communication refers to the use of untethered communication (e.g., infrared, acoustic, or radio frequency signals) for sending and receiving data between devices equipped with wireless interfaces [1]. Since its introduction, wireless communication has been a revolution for communication and networking technologies with the great advantages that it provides in comparison to its wired counterparts. With no wires needed for end-to-end communication, wireless

S. Abdel Hamid et al., *Routing for Wireless Multi-Hop Networks*,
SpringerBriefs in Computer Science, DOI: 10.1007/978-1-4614-6357-3_1,
© The Author(s) 2013

communication provides flexible deployment and use, cost reduction, mobility, network scaling, and convenience for both users and service providers. Having such advantages comes at the price of some drawbacks and limitations due to the characteristics of wireless communication, among them:

- Interference between wireless devices due to the broadcast nature of wireless communication; higher interference usually results in lower reliability of data transmission.
- Lower bandwidth and data rates compared to wired communication which results in higher delay/jitter and longer connection setup.
- Highly dynamic network conditions due to interference, loss of signal power with distance, and freedom of mobility [1].
- Fading due to obstacles and the "multipath effect".
- Frequency reuse due to limitation of bandwidth and spectrum.

One of the most prominent drawbacks is the limitation in transmission range due to the previously mentioned characteristics. Therefore, nodes using wireless communication can only communicate directly with nodes within their transmission range. There are two wireless communication setups sharing this feature:

1. Infrastructure-Based, where nodes communicate directly with an access point or base station that coordinates communication with other nodes,
2. Infrastructure-less or Ad-Hoc, where nodes communicate with one another without the aid of any coordinator or central controller.

In the ad-hoc mode, with limited transmission range, for a node to be able to communicate with another node out of its transmission range, it should depend on other intermediate nodes for packets to reach the intended destination. These intermediate nodes act as relays for the packet. This communication paradigm is known as "multi-hop" communication where a node can act as a source, a destination, or a relay.

A network comprised of nodes that use wireless multi-hopping for data transmission is known as a "wireless multi-hop network." A wireless multi-hop network suffers from all the previously mentioned drawbacks of wireless communication. Such drawbacks should be taken into consideration in designing an application or a protocol for a wireless multi-hop network and they are common for all the wireless multi-hop network paradigms.

As well, as a common characteristic for all wireless multi-hop networks, these networks are ad-hoc networks. This means that nodes in a network have no central control and they are responsible for cooperating with one another for handling network organization and management. Thus, wireless multi-hop networks are known to be self-organizing and self-configuring networks.

With the multi-hopping feature, wireless multi-hop networks have some unique characteristics and challenges in each layer of the protocol stack as well. For instance, signal interference and attenuation are major problems that need to be handled by the physical layer. Wireless nodes also suffer from the hidden and exposed terminal problems which should be handled by the deployed MAC protocol. The network layer faces great challenges with the most crucial one the task of establishing the communication path with its multi-hopping requirements, and this is the responsibility of the routing sub-layer.

As an essential part of the network layer, routing is a critical element in the design of networks and networking-based applications. In a multi-hop network, for a packet to be sent from a source to a destination, it is the responsibility of the routing protocol to find a path between the two communicating nodes through intermediate nodes that act as relays for packets. Having multiple intermediate nodes results in having multiple potential paths to be followed. So, the role of a multi-hop routing protocol is not only finding a path, but finding the optimal one that satisfies the needed performance requirements from a set of candidate paths.

Within the paradigm of multi-hop communication, four types of wireless networks can be classified as wireless multi-hop networks. These network paradigms are: Mobile Ad-Hoc Networks (MANETs) [1, 2], Wireless Sensor Networks (WSNs) [3, 4], Wireless Mesh Networks (WMNs) [5, 6], and Vehicular Ad-Hoc Networks (VANETs) [7, 8]. Being multi-hop networks, these paradigms depend on a sequence of intermediate nodes for routing packets from a source to a destination. Although having this feature in common, routing characteristics and functionalities of these four network paradigms have some differences that result in having different routing protocols uniquely designed for each network paradigm. These differences emerge from having different characteristics and challenges that impose requirements on the routing functions for each paradigm. They all utilize multi-hopping but with different techniques for handling the different routing components.

The objective of this brief is to highlight the commonalities and distinguishing features of the aforementioned network paradigms in terms of their routing functions. The remainder of this brief is organized as follows: in the following sections of this chapter, we will present a brief introduction to each of the aforementioned network paradigms. In Chap. 2, we will discuss the unifying features and the basic routing components that the paradigms have in common. As well, some auxiliary components will be discussed. In addition, we will propose a generic routing model that can be a foundation for the wireless multi-hop routing function and can be inherited by any wireless multi-hop routing protocol. In Chap. 3 we will shed light on the distinguishing features of each network paradigm in terms of their design considerations and challenges, and their effects on the routing functions. Furthermore, we will provide an abstraction of the general routing functionalities for each of the four network paradigms along with some examples of routing algorithms that adopt these functionalities. Finally, in Chap. 4 we will present open issues in the area of routing for wireless multi-hop networks, along with our view of the ideal wireless multi-hop routing protocol, and some concluding remarks.

1.2 Mobile Ad-Hoc Networks

A Mobile Ad-hoc NETwork (MANET) is a collection of mobile nodes with wireless networking interfaces that form a temporary network without the aid of any infrastructure or central control [1, 9]. Examples of these mobile nodes include laptops, notebooks, cell phones, and tablets. A simple architecture of a MANET is shown in Fig. 1.1. Nodes in a MANET are autonomous, self-organizing, self-configuring nodes that communicate in a multi-hop fashion and can move arbitrarily. Therefore, the network may experience rapid and unpredictable topology changes. The nodes in the network not only act as hosts but also as routers that route data to/from other nodes in the network.

MANETs originated from the Defense Advanced Research Projects Agency (DARPA) [10] Packet Radio Networks (PRENET) in the early 1970s. Being free of infrastructure, MANETs have many advantages such as ease of deployment, low cost, and high flexibility. Having these advantages, MANETs are appropriate for many commercial and industrial applications; for example, educational and file sharing purposes in conferences/meetings/lectures, emergency services, law enforcement operations, and home networking.

MANETs are considered the oldest wireless multi-hop network paradigm, and the other multi-hop paradigms can be considered as special cases of MANETs with some unique characteristics, application domains, and design requirements.

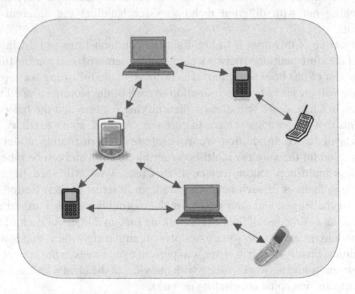

Fig. 1.1 A simple mobile ad-hoc network architecture

1.3 Wireless Sensor Networks

A typical Wireless Sensor Network (WSN) consists of a large number of small, inexpensive, resource constrained sensor nodes that communicate wirelessly in a multi-hop network. These sensor nodes collaborate together to accomplish a common sensing task and serve a certain application; for example, environment monitoring, battlefield surveillance, Intelligent Transportation Systems (ITS), home applications for domestic devices and user interaction, and industrial process control [3].

The sensor node basically consists of a sensing circuitry, a low-power embedded processor, a small memory, a radio transceiver, and a power source (usually a small battery). There are other optional components that are application-dependent, such as a Global Positioning System (GPS), a mobilizer, or a power harvesting system.

The sensing circuitry measures parameters from the surrounding environment and transforms them into electrical signals. These signals are processed by the node for analysis and decision making purposes. The sensor node sends such sensed data, usually via a radio transmitter, with the aid of other nodes in the network through multi-hopping, to a data-collection station (a base station or a sink) that may be connected to a command center, either directly or through the internet. A basic architecture of a WSN is shown in Fig. 1.2.

Usually, sensor nodes do not have energy feeding components except the small battery attached to the node; therefore, it is imperative for sensor nodes to survive with the very limited power supply they have. In some cases, energy harvesting techniques may be adopted; however, this is rarely used in order to keep the network deployment and operation costs as low as possible.

The most well-known sensor nodes are those developed by researchers at Berkeley and are known as "motes." Motes were made commercially available, along with TinyOS [11], an associated embedded operating system that facilitates the use of these devices.

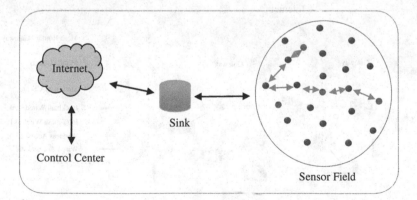

Fig. 1.2 Basic wireless sensor network architecture

1.4 Wireless Mesh Networks

Wireless Mesh Networks (WMNs) were proposed as an efficient technology to provide broadband access to users not in the direct coverage of wired access points by extending the backhaul access using wireless communications. Isolated Local Area Networks (LANs) can be connected together and coverage can be extended without incurring the cost and inconvenience of deploying a wired infrastructure.

A WMN is comprised of three tiers of networking components: (1) the mesh clients which are the user devices seeking access to the broadband network, (2) the wireless mesh routers (WMRs) that provide connectivity to the mesh clients and are connected together in a multi-hop fashion for covering the access area, and (3) the gateways which are connected to the mesh routers and provide the last mile access to the Internet. The general architecture is illustrated in Fig. 1.3. Only gateways need wired connection to the backhaul network. Being deployed in an ad-hoc fashion, wireless routers can be incrementally added to the network for further extension of the covered area as needed [5]. Except for the intra-mesh links which should be wireless, all other links can be either wired or wireless [12]. Clients can connect to the WMRs using any common network interface (e.g., 802.11, Ethernet, Bluetooth). WMRs can have multiple radio interfaces and can support multi-channel operation.

The extended coverage capabilities provided by WMNs motivates a promising market and applications. The main objective for designing WMNs was to provide public Internet access for areas not covered by wired infrastructure. In addition to its main objective, WMNs can support other applications: communications for ITS, public safety, broadband home networking, community and neighbourhood networking, enterprise networking and building automation [5].

Communication in WMNs can be in one of two patterns: communication between a mesh client and a gateway for the broadband access, or communication

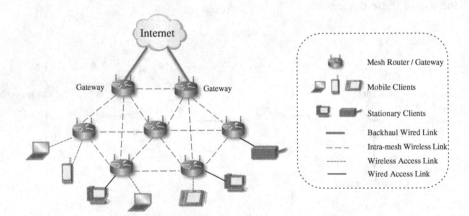

Fig. 1.3 Architecture and components of a wireless mesh network

between two mesh clients. In both cases, all transmissions are done via WMRs in a multi-hop fashion.

1.5 Vehicular Ad-Hoc Networks

A Vehicular Ad-Hoc Network (VANET) is the newest paradigm of wireless multi-hop networks. It emerged from MANETs with the mobile nodes being the vehicles on the roads. Vehicles communicate using wireless communication in a multi-hop fashion for disseminating information. These connected vehicles are known as intelligent vehicles and are equipped with a wireless communication module and sensors that monitor the interior and exterior surroundings and provide assistance/alerts to the driver via an on-board unit (OBU) [13].

Many standards have been introduced for VANET wireless communication with the most dominant being the Wireless Access for Vehicular Environment (WAVE) standard. WAVE is an amendment of the IEEE 802.11 standard for WLAN and it is standardized to be known as IEEE 802.11p [14].

Vehicles communicate with one another and with Road Side Units (RSUs) for relaying and sharing messages and information that will support many ITS application domains such as safety applications (e.g., broadcasting safety warnings), traffic management applications, road condition monitoring applications, infotainment applications (e.g., Internet access), advanced driver assistance services (ADAS) applications (e.g., automatic toll collection, remote diagnostics), to name a few.

Four communication patterns are available in VANET communications: (1) beaconing; 1-hop broadcasting for position and velocity information, (2) geocasting; sending information to an area of interest, (3) unicasting; sending information to a specific destination, and (4) information dissemination; flooding the surrounding area with information [15]. Figure 1.4 illustrates these communication patterns.

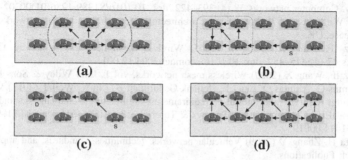

(a)   (b)

(c)   (d)

Fig. 1.4 VANET communication patterns: **a** Beaconing, **b** Geocasting, **c** Unicasting, and **d** Information dissemination

Among these communication patterns, geocasting, unicasting, and information dissemination are based on multi-hopping for delivering information. Both geocasting and unicasting require establishing a routing path between the source and the destination (which may be a specific area or node). Information dissemination may not need finding a definitive route but needs some routing functionalities for handling data redundancy and broadcast storms.

1.6 Summary

In this chapter, we presented an overview about wireless multi-hop networks and some general characteristics of their wireless communication. As well, we presented a brief introduction to the four existing wireless multi-hop network paradigms: mobile ad-hoc networks, wireless sensor networks, wireless mesh networks, and vehicular ad-hoc networks.

Many factors can be used to profile the four multi-hop network paradigms. Some prominent factors are network size, terrain scope, mobility degree, rate of topology change, type of traffic, QoS, energy constraints, availability of computational resources, location dependency, and addressing schemes. We will elaborate on the features of each network paradigm in terms of these factors in Chap. 3 where we will present detailed discussions on the characteristics and the distinguishing features of each network paradigm and how these features will affect the routing function of each paradigm.

References

1. Basagni S, Conti M, Giordano S, Stojmenović I (2004) Mobile ad hoc networking. Wiley-IEEE Press, USA.
2. Boukerche A (2009) Algorithms and protocols for wireless and mobile ad hoc networks. Wiley, USA.
3. Akyildiz IF, Su W, Sankarasubramaniam Y, Cayirci E (2002) Wireless sensor networks: a survey. Computer networks 38 (4):393–422. doi: 10.1016/S1389-1286(01)00302-4.
4. Karl H, Willig A (2007) Protocols and architectures for wireless sensor networks. Wiley-Interscience,.UK.
5. Akyildiz IF, Wang X, Wang W (2005) Wireless mesh networks: a survey. Computer networks 47 (4):445–487. doi: 10.1016/j.comnet.2004.12.001.
6. Akyildiz IF, Wang X (2009) Wireless mesh networks, vol 1. John Wiley & Sons Inc, UK.
7. Karagiannis G, Altintas O, Ekici E, Heijenk G, Jarupan B, Lin K, Weil T (2011) Vehicular networking: A survey and tutorial on requirements, architectures, challenges, standards and solutions. IEEE Communications Surveys & Tutorials 13 (4):584–616. doi: 10.1109/SURV.2011.061411.00019.
8. Moustafa H, Zhang Y (2009) Vehicular networks: techniques, standards, and applications. Auerbach Publications.
9. Johnson DB, Maltz DA (1996) Dynamic source routing in ad hoc wireless networks. Mobile computing 353:153–181.

10. Jubin J, Tornow JD (1987) The DARPA packet radio network protocols. Proceedings of the IEEE 75 (1):21–32. doi: 10.1109/PROC.1987.13702.
11. Hill J, Szewczyk R, Woo A, Hollar S, Culler D, Pister K (2000) System architecture directions for networked sensors. ACM Sigplan Notices 35 (11):93–104. doi: 10.1145/356989.356998.
12. Jun J, Sichitiu ML (2008) MRP: Wireless mesh networks routing protocol. Computer Communications 31 (7):1413–1435. doi: 10.1016/j.comcom.2008.01.038.
13. Hartenstein H, Laberteaux KP (2008) A tutorial survey on vehicular ad hoc networks. IEEE Communications Magazine 46 (6):164–171. doi: 10.1109/MCOM.2008.4539481.
14. Jiang D, Delgrossi L (2008) IEEE 802.11 p: Towards an international standard for wireless access in vehicular environments. In: IEEE Vehicular Technology Conference, 2008, VTC Spring '08. pp 2036–2040. doi: 10.1109/VETECS.2008.458.
15. Schoch E, Kargl F, Weber M, Leinmuller T (2008) Communication patterns in VANETs. IEEE Communications Magazine 46 (11):119–125. doi: 10.1109/MCOM.2008.4689254.

Chapter 2
Routing for Wireless Multi-Hop Networks: Unifying Features

Abstract Wireless multi-hop networks share some routing features based on the fact that they all follow the multi-hopping paradigm. In this chapter, we follow a component-based approach for breaking down a routing protocol into some core and auxiliary components. We discuss the core components that are fundamental for any wireless multi-hop routing protocol along with some auxiliary components that can be adopted to achieve a specific design goal. Dependency and relationships among the components are elaborated as well. Finally, we propose a generic routing model that can be inherited for the design of any wireless multi-hop routing protocol.

Keywords Wireless multi-hop networks · Route discovery · Route selection · Route representation · Data forwarding · Route maintenance · Route energy efficiency

2.1 Introduction

Routing is the main function of the network layer, the 3rd layer of the protocol stack, and its performance is highly affected by the lower layers: the physical and data link layers. In order for a routing protocol to be efficient and reliable, the protocol designer should consider the effects of the lower layers and provide mechanisms for handling these effects. For example, due to some features of the physical layer, the communication range of the devices/communicating nodes may be asymmetric. This means that if node A can send a message directly to node B, it is possible that node B cannot reply back directly to node A. These communication issues should be taken into account when designing a routing protocol. On the other hand, this kind of cross layer effect can be utilized to improve the performance of the routing protocol. For example, routing protocols that are designed to support QoS and low latency requirements must consider link qualities in choosing the optimal path among the set of available paths. These link quality measures can be obtained from the lower layers by passing parameters to the network layer. A designer of a routing protocol then should consider the functionalities of the lower layers, handle their affecting features, and utilize their measures, parameters, and, in some cases, their layer-specific packets.

For the wireless multi-hop networks, MANETs, WSNs, WMNs, and VANETS, there is no single wireless multi-hop routing protocol which can fit all needs. This is because each network paradigm has its own design challenges. Yet, as they all are classified under the category of wireless multi-hop networks, they have some unifying features. There are some routing functionalities and components that are essential, and are common parts of any wireless multi-hop routing protocol.

In Ref. [1], Lee et al. proposed a taxonomy that can be followed in designing a wireless routing protocol. They propose breaking down the wireless routing protocol and functions into multiple smaller components. Some of these components are core ones that should be a part of any wireless routing protocol and others are auxiliary that can be included only when needed by the application requirements. Following this component approach, in the following section, we will provide a detailed discussion of the routing components showing the core and auxiliary ones, and when these auxiliary ones may be needed.

2.2 Routing Components: An Exhaustive View

By breaking down the wireless routing protocol into smaller components, we can analyze the components that should be included in any wireless multi-hop routing protocol and show the interacting behavior between them. The behavior of these basic components can be tailored to different application profiles and needs, while keeping and maintaining the core functional behavior and goals [1]. To satisfy network and application specific needs, extra components can be added to the routing protocol to control its behavior and maintain its performance as needed and specified by the application and network paradigm. Having the core components, a routing protocol can be easily extended to accommodate and support extra requirements, services and features by adding auxiliary components. In the two following sub-sections, we will discuss the core components that should be a part of the skeleton of any routing protocol and we will shed light on some auxiliary components that may be used only based on the network and application needs.

2.2.1 Core Components

These components are considered to be basic building blocks for any wireless routing protocol to provide its main function, getting a message from a source to a destination. These components are route discovery, route selection, and route representation and data forwarding.

2.2.1.1 Route Discovery

Route discovery is the first stage of the function of any wireless routing protocol. Route discovery is the process of finding a route/set of potential routes between a source and an intended destination. The process of finding a route can be classified into three categories: *proactive*, *reactive* or *hybrid*.

Proactive route discovery, also known as *table-driven* route discovery, depends on the use of up-to-date routing information about the whole network to find a path from any source to any destination in the network. This routing information is exchanged among nodes either periodically or upon the occurrence of any change in the network topology. This information is kept at each node in a routing table. This type of route discovery pre-determines routes between any two nodes irrespective of the need for such routes. When a node has a packet to be sent, it does not need to wait for a route to be discovered. It consults its routing table, gets the up-to-date recorded route, then sends the packet without incurring a delay for the route to be discovered—the route is discovered a priori.

There are two sub-categories under the proactive routing category: Distance Vector (DV) and Link State (LS). They differ in how the network topology information is spread. These techniques are borrowed from wired networks but they can be modified to handle the characteristics of MANETs.

(a) **Distance Vector Proactive Routing**

In DV route discovery, each node maintains a routing table where it stores information about all possible destinations, the next node to reach that destination, and the best known distance to reach the destination.[1]

These tables are updated by exchanging information with the neighbors. Each node periodically sends a vector to its direct neighbors carrying the information recorded in the routing table to maintain topology. The distance vector contains the destinations list and the cost—the distance—to reach each destination.

The basic distance vector routing technique works in theory but has a serious drawback in practice. It suffers from a severe problem known as "count-to-infinity" [2]. This happens as a result of the occurrence of routing loops; when X tells Y that it has a path somewhere, Y has no way of knowing

---

[1] Distance can be defined as the number of hops.

whether it itself is on the path. This drawback is common in the DV routing technique, and all DV-based routing protocol designers should consider this issue and find a mechanism to avoid it.

(b) **Link State Proactive Routing**

Distance vector routing was used in ARPANET until 1979, when it was replaced by link state routing. The objective of LS routing is to provide an alternative to DV that avoids routing loops and the subsequent "count-to-infinity" problem. LS routing overcomes this by maintaining global network topology information at each node.

In LS routing, each node periodically sends information about the cost to reach each of its direct neighbors and it includes this information in what is known as the link state packet. This link state packet is sent to all the other nodes in the network by flooding. Each node does the same link state flooding procedure and, eventually, each node will have link state packets from all other nodes, so each node will have information about the complete topology and costs of all the links in the network. Then Dijkstra's algorithm [3] can be run locally to construct the shortest path to all possible destinations. The results of this algorithm can be stored in the routing tables for later use [2].

Although LS routing avoids some problems with DV routing, it has a problem with its storage requirements.

As an advantage, proactive route discovery incurs almost no delay as routes are calculated in advance and are available in the routing table. However, it has a disadvantage that may hamper its use in large networks. It incurs an overhead related to the periodic routing updates which may cause congestion for the network when it has a large number of nodes. Therefore, in most cases, the proactive route discovery has problems with network scalability.

Reactive route discovery is also known as *on-demand* route discovery. As the name implies, the route is discovered on demand. When a source has a packet to be sent, it initiates a route discovery process to set up a path to the intended destination. Many approaches can be followed for path setup where the most common one is having the source node broadcast a route request packet carrying the destination address and asking for a route to that destination. When the route request reaches the destination or an intermediate node that knows a route to that destination, a route reply packet is sent back to the source carrying details about the discovered route.

Some protocols perform route discovery on the fly, hop-by-hop. When a node receives a packet to be forwarded to another node, it decides to which neighbor it should forward this packet. This type of routing is known as self-routing and it falls under the category of reactive routing as the route is established on demand. An example of this type of routing is geographical routing where a node picks the next hop based on the locations of its neighbors and their distances to the destination. The self-routing based protocols usually require a form of neighbor discovery to know about the potential forwarding nodes that the current node will choose from to be the next hop.

Reactive route discovery avoids the drawback of the proactive one by avoiding exchanging periodic routing updates, which reduces the traffic overhead. However, as the path is discovered only on-demand, this type of route discovery incurs a delay overhead and longer latency for route establishment.

The category of *hybrid* route discovery is obtained by combining both the proactive and reactive techniques to make use of the advantages of both and mitigate their disadvantages. It tries to reduce the control overhead associated with proactive route discovery and the delay incurred in the reactive one.

2.2.1.2   Route Selection

As an output of the route discovery stage, there will be a set of potential routes between a source and destination. It is the role of the route selection component to pick the optimal path from this set that satisfies the needed performance criteria. Most of the routing protocols are based on choosing only one path for delivering packets from a specific source to a specific destination; however, there are some protocols that rely on choosing multiple paths (multipath routing) [4] to provide redundancy and fault tolerance for the routing process.

For the proactive protocols, route selection is done implicitly with the route discovery stage. When the network topology information is shared and received by the nodes, they update the information in their routing tables accordingly; hence, routes available in the routing tables are the selected, best ones at that time.

Route selection in the reactive protocols is a stand-alone process. It can be handled by the source, the destination, or the intermediate nodes. In *destination-based* route selection, when the destination receives multiple route requests forwarded by multiple intermediate nodes, it can select the path to receive data through and sends the route reply along this path. The destination can pick the first path through which it received the first route request, the fastest one, or it can wait for a specific period of time. Thereafter, if it has received many route requests, it can pick the optimal path according to some selection metrics, discussed later in this section.

In *source-based* route selection, the source node may receive multiple route replies from the destination,[2] or from all intermediate nodes that know about a route to the intended destination. It is the responsibility of the source to pick a route from the set of routes extracted from the multiple route replies.

For *intermediate-based* route selection, the intermediate nodes decide on which route a packet should follow to reach a destination. They can either choose a route from a set of possible routes they keep for that destination,[3] or select a next hop on the fly. This type of route selection is involved in self-routing protocols. Since the

---

[2] The destination may reply to all route requests it receives.

[3] These routes may be discovered by them in a previous interaction with the destination or overheard from neighbors interacting with that destination.

route is selected on a hop-by-hop basis, the intermediate nodes are involved in the route selection process when the packet is forwarded to any of them.

Whether route selection is done by source, destination, or intermediate nodes, the deciding node should depend on one or several metrics for the selection decision. Determination of which route metric to use is dependent on the application requirements and needs. The simplest route metric and the most popular one is the hop count. The path with the least hop count will be chosen to reduce the number of intermediate nodes involved in the routing process and so reduce the control overhead and contention level among nodes. Examples of other routing metrics include energy consumption level, residual energy of the next hop, QoS metrics (such as end-to-end delay/jitter, interference level, packet loss rate, link residual capacity, and load balancing), link security level, and memory consumption. Some of these routing metrics require parameters related to the lower layers like the QoS-based link quality ones. These parameters can either be passed from the lower layers to the routing layer, or, in some protocols, this interaction with the lower layers is done in the form of cross-layer protocol design.

In short, how the route is selected is based on the application/network paradigm for which the protocol will be used. It is how the route will be selected that controls the performance of the routing protocol and whether it will satisfy the needs of the application or not.

2.2.1.3 Route Representation and Data Forwarding

After selecting a route, it should be stored to be followed for data transfer. We consider both route representation and data forwarding as a single component as they are highly integrated together and, in many protocols, they are done simultaneously. Route representation and data forwarding can follow one of two techniques: *exact route* and *route guidance* [1].

(a) **Exact Route**
   In this technique, the sequence of intermediate nodes that a path should follow to reach a destination is represented explicitly. There are two approaches for using the exact route representation and forwarding. These approaches are routing table and source routing.

   • *Routing Table*
     In this approach, each node keeps a routing table where it stores the next hop to reach potential destinations with one entry per destination. In the proactive protocols, this routing table contains information and next hops to all other potential destinations in the network. In the reactive protocols that make use of the routing table approach, they keep information about the destinations that they interacted with previously or those nodes that they overheard paths to them. Also, in these routing tables, they may keep information about nodes from which they received route requests or route replies for further relaying. When a packet is to be forwarded, the node looks up the routing table and gets

the next hop to which it should forward the packet to reach the intended destination.

- *Source Routing*

  The idea of the source routing approach is to include the whole path that the packet should follow in the packet header, and when a node gets this packet, it can extract the next hop from the path included in the packet. This path is included by the source node before sending the packet. This approach encounters some problems especially with large networks; as the complete path is included in the packet header, this can be considered traffic overhead and a source for bandwidth wasting.

(b) **Route Guidance**

In route guidance-based protocols, the sequence of intermediate nodes is not explicitly described. The full path is not determined prior to sending the packet by the source, rather the path is formed on the fly (i.e., self-routing). As the route is not fully determined a priori, nodes cannot store information about the path itself but they may store information about how the next hop will be chosen or information that will be used for picking the next hop. This is what is called route guidance. The geographical routing protocols are examples of protocols that follow the route guidance technique. In these protocols, instead of keeping information about the path itself, nodes store the positions of their neighbors and pick the next hop on the fly based on the destination and their direct neighbors' positions [5].

The three aforementioned components are considered core ones that should be included in any wireless routing protocol. As mentioned above, their behavior can be tailored to meet the requirements of the network paradigm that the protocol is designed for and this will be discussed in Chap. 3. In the following section, we will explore some of the auxiliary routing components that can be added to the core components to achieve a specific design goal.

2.2.2 Auxiliary Components

These components are not essential for all routing protocols but they can be added to improve the performance of a protocol or to make it meet the requirements and needs of a specific application or network paradigm. Examples of these components are route maintenance, route energy efficiency, and route security. Some of these components are discussed in the following.

2.2.2.1 Route Maintenance

The goal of the route maintenance component is to keep a route valid while in use and to handle possible failures. Route maintenance is needed by networks where links are prone to failure due to node mobility, for example. It is considered a crucial component in MANETs where nodes are highly mobile and the network topology encounters frequent dynamic changes. Route maintenance includes *route refreshing*, *route failure handling*, and *route invalidation* [1].

(a) **Route Refreshing**

Route refreshing aims at keeping the current routes valid by updating or using them only for the sake of refreshing. Route refreshing can be handled by one of three approaches depending on the category of the protocol: proactive, reactive, or hybrid. In the proactive protocols, route refreshing is done implicitly by having the nodes periodically or upon the occurrence of topology changes exchange network topology information and update the routing tables according to the current changes in the network. Therefore, in the proactive protocols, routes in the routing table are always the most up-to-date. In the reactive approach, routes are only touched on demand, so to keep routes usable and ensure their validity, nodes can refresh routes either by use of control packets (e.g., hello messages) or by using a data packet before the expiration of the route. Hybrid protocols and hybrid route refreshing combine both the proactive and reactive approaches.

(b) **Route Failure Handling**

In reactive routing, when an intermediate node finds that the next hop is unreachable, it tries one of two options: (1) to find an alternate path locally either by looking up its routing cache for an alternative or by initiating a route discovery process to replace the failing link with a valid one, or (2) to send a route error message to the source node with information about the failing link. The source node can also look up its route cache for a different route. If there is no alternative, it reinitiates the route discovery process while marking the failing part in order not to include it again.

In proactive routing, route failure is handled by route refreshing. As the routing tables have up-to-date routing information, route failure is handled by automatic updates.

In the hybrid protocols, it is a combination between the proactive and reactive route failure handling approaches.

(c) **Route Invalidation**

Route invalidation is the process of finding out stale routes and removing them from the routing tables and caches. The stale routes are distinguished and recognized by employing a lifetime period for each route, and if this route has not been refreshed during that period, it will be marked as expired and will be removed.

2.2.2.2 Route Energy Efficiency

As some of the wireless multi-hop networks are comprised of devices with limited resources, e.g., sensor nodes in WSNs, such networks have energy efficiency as one of the major design considerations that should be taken care of in any protocol designed for such networks including the routing ones. Routing protocols designed for such networks should include mechanisms to conserve node energy to prolong the lifetime of the nodes and of the network as a whole. Examples of such techniques are data aggregation, use of meta-data, load balancing, restricted flooding, use of energy-aware metrics, use of a resource manager, and putting nodes into sleep mode.

(a) **Data Aggregation**

Data aggregation is one of the techniques that is highly utilized in the energy-efficient routing protocols because, when deployed, it has a great impact on the nodes' residual energy and lifetime. The idea is that instead of sending redundant packets or packets that have a kind of correlation, these packets can be combined and aggregated together into only one packet. Reducing the number of transmitted packets leads to great conservation in node energy.

(b) **Use of Meta-Data**

A number of protocols depend on sending meta-data that describes the actual data packets instead of sending the actual packets themselves. This technique is mainly used for advertising the actual data. Instead of sending long data packets to nodes that may not be interested in them, small meta-data is sent to advertise the acquired data packets and if a node shows its interest in such data, the complete data packet is sent to it afterwards.

(c) **Load Balancing**

Many protocols focus on balancing the traffic load among the nodes in order not to overload some nodes compared to others which may lead to depletion of these nodes' batteries and cause their failures. For example, in cluster-based routing protocols, if cluster formation is static and not changed throughout the network life, the nodes that act as cluster-heads will burn their energy quickly, and after they die, all their members will be "headless" and therefore useless. This is because the role of being a cluster-head is energy consuming as the cluster-head has to be awake all the time, receive data from all of its cluster members, incur processing overhead for aggregating the data, and is responsible for the long-range transmissions to the data collector. To provide energy efficiency and balance energy consumption among the nodes, some routing protocols utilize dynamic clustering to rotate the role of being a cluster-head among the nodes.

(d) **Restricted Flooding**

When a packet needs to be broadcast (e.g., route request packets or data interests), some protocols make use of restricted flooding instead of flooding the packet to the whole network. For example, the packet can be sent to a group of nodes with higher probability to forward the packet or with wider coverage and view for the network. Another example is forwarding the packet to an area of interest instead of to the whole network, for example, sending data interests geographically to the area of interest then flooding the interest only within this area.

(e) **Use of Energy-Aware Metrics**

When this technique is utilized, it can be considered a part of the route selection component. To conserve energy, the optimal route can be selected based on the energy of the nodes constituting that route. A node's current energy consumption level or current residual energy can be used as the route selection metric.

(f) **Use of a Resource Manager**

Some protocols add to the routing component a resource manager that monitors the energy level of the nodes and adjusts their operations based on some thresholds.

(g) **Putting Nodes into Sleep Mode**

As a common technique in most of the WSN protocols (either MAC, routing, or other layer protocols), putting nodes into sleep mode saves a significant amount of energy. In the sleep mode, only the processor works with only a small portion of its capabilities; neither sensing nor transmissions are done. Once the node gets tasked or awakened, it works with all its capabilities.

2.3 Generic Routing Model

In this section, we will present a generic routing model that can be used to form the foundation of a wireless multi-hop routing protocol. We will present the functionalities as blocks and methods that can be selectively utilized and combined together to form a wireless routing protocol suitable for any wireless multi-hop network. This generic model can be further extended and enhanced with auxiliary functionalities to meet specific requirements per network paradigm.

Each component will be presented with its own various functionalities that will be available to the protocol designer to choose from. The output and the input of each component will be shown to clarify the interactions between the various components. The proposed generic model is shown in Fig. 2.1.

The route discovery component has five options/functions for the designer to choose from: (1) proactive with distance vector, (2) proactive with link state, (3) reactive with deterministic routing, (4) reactive with self-routing (which requires that each node discovers its neighbors; therefore, it calls the neighbor discovery function which feeds it with the neighbors list), and (5) hybrid discovery.

The route selection component has three functions for the protocol designer to choose from: (1) source-based selection, (2) destination-based selection, and (3) intermediate-based selection. The choice of which function to be used depends on the route discovery function that has been chosen (e.g., the reactive self-routing discovery requires the use of intermediate-based route selection).

Finally, the route representation and data forwarding component has three functions available for the designer's choice: (1) representation and forwarding

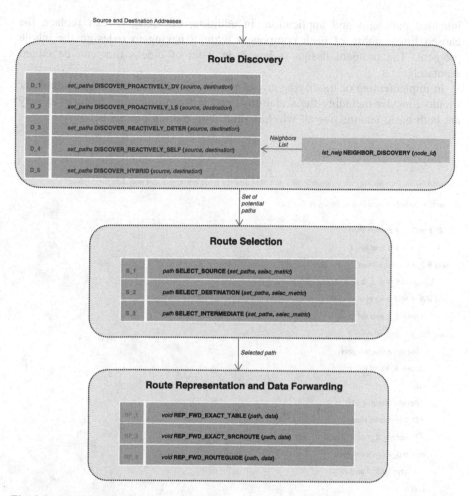

Fig. 2.1 Generic routing model

using exact route with routing tables, (2) representation and forwarding using exact route with source routing, and (3) representation and forwarding using route guidance. Again, the choice of the appropriate function strictly depends on the chosen discovery function (e.g., the reactive self-routing discovery requires the use of route guidance).

The following pseudo-code shows the interaction and dependency of the route selection function and the route representation and data forwarding function to be chosen and the already chosen discovery function. For simplicity, we refer to the functions by codes—these codes are shown in Fig. 2.1 next to their associated functions.

By breaking down the functionalities into blocks and methods, the protocol designer can choose whatever functionalities are preferred and suitable for the intended paradigm and application. In addition, the designer can replace the chosen functionality of each component without having to redesign the whole protocol. The protocol design is based on a set of blocks that can be edited separately.

In implementing or modifying any of the functionalities, the protocol designer should consider including the scalability[4] and self-configuration[5] features as they are both basic features for all wireless multi-hop routing protocols.

Pseudo -code for choosing the route selection function and the route representation and data forwarding

function based on the chosen discovery function

```
if D_1 or D_2 is chosen then
        choose S_1 and RF_1
else if D_3 is chosen then
        choose S_1 or S_2 or S_3  and  RF_1 or RF_2
else if D_4 is chosen then
        choose S_3 and RF_3
else if D_5 is chosen then
        // For the proactive part
        choose S_1 and RF_1
        and
        // For the reactive part
        if D_3 is chosen then
                choose S_1 or S_2 or S_3  and  RF_1 or RF_2
        else if D_4 is chosen then
                choose S_3 and RF_3
        end if
end if
```

2.4 Summary

In this chapter, we discussed the unification of multi-hop networks in terms of their routing functions. We followed a component-based approach for breaking down a routing protocol into some core and auxiliary components. We presented the core components that are considered a part of any wireless multi-hop routing protocol and are considered the common and unifying features of all wireless multi-hop

---

[4] Since the network is ad hoc, the number of nodes can always increase.

[5] Note that there is no central control.

routing schemes. As well, we discussed some auxiliary components that can be added to the core ones to achieve a certain design goal. Finally, we introduced a generic routing model that can be inherited by, and considered the basis of, any wireless multi-hop routing protocol.

References

1. Lee MJ, Zheng J, Hu X, Juan H, Zhu C, Liu Y, Yoon JS, Saadawi TN (2006) A new taxonomy of routing algorithms for wireless mobile ad hoc networks: the component approach. IEEE Communications Magazine 44 (11):116–123. doi:10.1109/MCOM.2006.248174
2. Tanenbaum AS (2003) Computer Networks. 4 edn. Prentice Hall,
3. Dijkstra EW (1959) A note on two problems in connexion with graphs. Numerische mathematik 1 (1):269–271. doi:10.1007/BF01386390
4. Mueller S, Tsang R, Ghosal D (2004) Multipath routing in mobile ad hoc networks: Issues and challenges. Performance Tools and Applications to Networked Systems:209–234
5. Mauve M, Widmer A, Hartenstein H (2001) A survey on position-based routing in mobile ad hoc networks. IEEE Network 15 (6):30–39. doi:10.1109/65.967595

Chapter 3
Routing for Wireless Multi-Hop Networks: Distinguishing Features

Abstract Although the wireless multi-hop network paradigms have some unifications in terms of their routing function, they have many distinguishing features based on the fact that each paradigm has its own characteristics and application demands. In this chapter, we explore the distinguishing features that shape the routing functionalities of each of these paradigms. We present a discussion of the design considerations of routing protocols for each paradigm, the popular classifications of such routing protocols, and the core routing components of each paradigm along with some routing functionalities of each component and some representative protocols that adopt these functionalities. Finally, we present a summary of the core routing components and main functionalities for each paradigm for the sake of a concise and clear comparison of the various wireless multi-hop network paradigms.

Keywords Wireless routing · Wireless multi-hop networks · Wireless ad-hoc networks · Wireless sensor networks · Wireless mesh networks · Vehicular ad-hoc networks

3.1 Introduction

As mentioned in Chap. 1, there are four wireless networking paradigms—MANETs, WSNs, WMNs, and VANETs—that can be classified as wireless multi-hop networks as they all follow the pattern of wireless multi-hopping in sending packets among their networked nodes. Although these network paradigms have many characteristics in common, as each consists of different types of nodes and serves different applications, each has its own unique features, requirements, and design challenges that necessitate the need for a set of protocols designed specifically

S. Abdel Hamid et al., *Routing for Wireless Multi-Hop Networks*,
SpringerBriefs in Computer Science, DOI: 10.1007/978-1-4614-6357-3_3,
© The Author(s) 2013

for each network paradigm. In this chapter, we will show the distinguishing characteristics and challenges for each of the four aforementioned network paradigms, the most popular classification for each paradigm, and the core components and distinguishing routing functionalities for each network paradigm.

For each of the following sections, we will present a discussion of the distinguishing features of all the wireless multi-hop network paradigms to serve as a comparison of the four paradigms and help the reader assimilate the distinctions of these paradigms in terms of their routing requirements and functionalities.

3.2 Distinguishing Design Considerations

In this section, we present the design considerations for each of the wireless multi-hop network paradigms that should be taken into account by a routing protocol designer to achieve the paradigm's specific needs and meet its functional requirements.

Mobile Ad-Hoc Networks

Routing is considered a vital component for the operation of a MANET due to the network dynamics, and many protocols have been proposed in the area of routing in MANETs. There are several design challenges that need to be considered in designing a routing protocol for a MANET:

(a) **Node Mobility**

As the nodes are free to move, the network topology keeps changing. In addition to being either a source or a destination, nodes in MANETs also serve as routers for other nodes' transmissions, so the routing paths are always prone to failure. It is the responsibility of the routing protocol, besides discovering the route, to provide a recovery from such path failures. The routing protocol should have a route maintenance component to ensure that packets reach their destination in the shortest time even if the original path has been broken for some reason. The frequent change of network topology is considered the most significant and critical challenge for routing in MANETs.

(b) **Resource Constraints**

Nodes in MANETs are often portable devices that are limited in their hardware resources. This imposes limitations on the complexity of the designed protocol.

(c) Scalability

The routing protocol should provide an acceptable level of service even in the presence of a large number of nodes. Newly added nodes should be identified rapidly. In addition, the control packets must be utilized efficiently to deliver data packets, and be generated only when necessary in order not to cause network implosion when having a large number of nodes with frequent movements.

(d) Lack of Infrastructure and Central Control

Being ad-hoc networks, MANETs should be self-organizing and configuring; there is no centralized administration entity to manage the operation of the different mobile nodes. The routing protocol should depend on distributed techniques for network configuration and assigning roles for hierarchical structures (e.g., clustering).

(e) Limited Bandwidth

As the nodes communicate using a wireless interface, the available bandwidth is limited. The routing protocol should enhance bandwidth utilization for better performance of the whole network. It should reduce the number of control packets as much as possible and minimize the data packet header size while keeping the protocol performance as efficient as possible.

All these design challenges should be taken into consideration in designing a routing protocol for MANETs with the main focus being on the most critical challenge: handling node mobility to maintain network connectivity.

Wireless Sensor Networks

Designing a routing protocol for WSNs is very challenging due to the WSN unique characteristics that distinguish this type of network from other wireless networks. The design challenges in sensor networks involve the following main aspects:

(a) Energy Constraints

Sensor nodes are restricted by their power supply as they are usually battery-powered, and in most applications it is difficult or infeasible to replace or recharge the batteries (e.g., in harsh environments). Having this power constraint, WSN applications and protocols should be designed to be as energy-efficient as possible to prolong both the node and network lifetimes.

(b) Limited Hardware Resources

The sensor node is also limited in its processing and storage capabilities, therefore WSN protocols should be as simple as possible and with low data storage requirements.

(c) **Dense Deployment**

Sensor nodes are often deployed in large numbers and this number can be several orders of magnitude higher than that in a MANET; there can be hundreds of thousands of nodes in very large areas. The WSN protocol should work efficiently with this number of nodes and try to reduce the communication and control overhead that may cause network congestion and performance degradation.

(d) **Addressing Scheme**

Due to the dense deployment in WSNs, it is not possible to build a global addressing scheme. Thus, the IP-based protocols may be not applicable to WSNs. In addition, in WSNs, nodes collaborate together to achieve the overall application goals and perform the sensing tasks. In most cases, there is no interest in which particular node reported the data—the interest is in the reported data itself and not in which node it was sent from. The common addressing scheme in WSNs is a data-centric scheme which is based on attribute-based addressing, where data is represented as an attribute-value pair that may be requested by queries sent by the base station or reported by the node in a time-based or event-based manner.

(e) **Scalability**

Nodes are deployed densely in a WSN and more nodes may be added during the network operation to provide more coverage or accuracy. The protocol should accommodate all these nodes and be scalable to different network sizes.

(f) **Self-Configuration**

Being an ad-hoc network, it is the responsibility of the nodes to configure themselves on the fly once deployed and to organize themselves into a communication network. The routing protocol should provide distributed techniques to support this feature.

(g) **Fault Tolerance**

Due to the limited resources, sensor nodes are prone to failure either through hardware failure or depleted batteries. This failure also may be due to environmental factors and unattended operation. The protocol for a WSN should be fault tolerant, handle frequent topology changes, and utilize self-repairing and self-recovery mechanisms.

(h) **Data Redundancy**

Due to the dense deployment of the sensor nodes, there are many nodes in an area of interest. The data sensed by these nodes are based on a common phenomenon and have some sort of correlation and redundancy. Exhausting the network with this redundant data causes problems with unneeded energy consumption and inefficient bandwidth utilization. On the other hand, this redundancy of data can be advantageous. By utilizing data aggregation techniques, this redundancy can be exploited by the routing protocol to decrease the number of transmissions and hence, improve the energy efficiency and the bandwidth utilization.

(i) Diverse Applications

WSNs have a wide range of applications each with its own requirements. Therefore, there is no "one-for-all" protocol; the requirements of the data gathering applications are different from those that focus on critical event reporting. The protocol designers should develop the protocol in a way that satisfies the needs of the application and utilizes the resources efficiently.

Having all these unique challenges for WSNs has forced the design of new routing protocols instead of using the existing protocols of MANETs. New protocols have been needed to meet all these design challenges and provide the needed network performance. To make the WSN operational for several years, the energy efficiency should be the primary design goal in any WSN protocol.

Wireless Mesh Networks

Routing for WMNs plays a crucial role for providing WMN services with the required quality for users. WMNs have some unique characteristics that impose distinguishing challenges, and features that offer opportunities for improvement when designing a WMN routing protocol. The most prominent ones are:

(a) QoS Guarantee

As the main objective of a WMN is providing broadband services, routing for WMNs should depend on the use of QoS and link quality metrics to satisfy the QoS requirements; minimize delay, provide real time communications, and balance the load between the multiple available paths.

(b) Robust Coverage

To guarantee QoS and provide efficient services, the WMN should be more robust to network faults and link failures. Dead zones can be eliminated by adding mesh routers and by the proper placement of routers to avoid service disruptions. A WMN routing protocol should utilize the availability of multiple paths between the potential source–destination pairs. Such redundancy of communication paths can also impose challenges for the routing protocol due to increased complexity for maintaining these paths.

(c) Minimal Mobility

In the backbone WMN, WMRs are almost stationary. With such static deployment, mobility support in WMNs is not a major concern. This releases the burden and the complexity of handling mobility by the routing protocol. Also, being static, WMRs can be easily hooked up to permanent power supplies; hence, also no power constraints need to be handled by the routing protocol.

(d) Multiple Radios

WMRs may utilize multiple radios to increase capacity and improve QoS.[1] This feature creates a unique challenge for the WMN routing protocols to handle. Also, having multiple interfaces requires cooperation between the routing layer and lower layers involving passing parameters among the layers or utilizing cross-layer functionalities.

(e) Adaptive Support for Both Mesh Routers and Clients

Most of the protocols are designed for the backbone mesh. New routing protocols are needed that include mechanisms for handling both the backbone WMN and client WMN features.

(f) Scalability

QoS guarantees can require keeping the end-to-end delay as low as possible. Setting up a path in a very large network may incur a long delay. The routing protocols should be designed to be scalable and include mechanisms for keeping the end-to-end delay reasonable for all network sizes.

These features and challenges uniquely distinguish the routing functions in WMNs. Some of them can be considered advantageous as they relax some constraints that are present in other network paradigms and others can be considered difficult challenges that should be handled in the routing protocol design.

Vehicular Ad-Hoc Networks

There were many trials for using MANET routing protocols with VANETs, based on the idea that a VANET is a special type of MANET. These trials did not achieve the expected performance as the VANET environment showed that it presents some unique challenges and features that entail the development of a set of new routing protocols or the adaptation of some current MANET protocols to meet the vehicular environment needs. Some of these distinguishing challenges and features are:

---

[1] Utilizing multiple radios enables separation of two main types of traffic; while routing and configuration are performed between mesh routers on one radio, the access to the network by end users can be carried out on a different radio.

(a) Highly Dynamic Topology

Since vehicles are moving at high speeds, especially on highways, the topology in a VANET is subject to frequent changes. Routing protocols should provide mechanisms for maintaining the followed routes and handling link changes.

(b) Intermittent Connectivity

It can be a common case in VANETs that a vehicle will not find a neighbor in its vicinity to forward its data to and it will have to keep the data till it comes into contact with another vehicle. This will be the case in sparse environments. Even in dense environments, traffic lights and stop signs may lead to some network partitions. Routing protocols should provide mechanisms for handling this intermittent connectivity. The most common mechanisms are the Store-Carry-Forward mechanism for delay-tolerant networks and either the use of Road Side Units (RSUs) for relaying messages, or depending on finding an alternative path using a recovery mechanism in delay-sensitive networks.

(c) Restricted Mobility Patterns

Vehicles mobility patterns are restricted by road topology and speed limits. This may be considered advantageous because these restricted patterns can help in predicting future conditions (e.g., traffic conditions and vehicles positions). This feature will help the routing protocols to make more informed decisions [1].

(d) Sufficient Resources

Vehicles have several advantages over other types of mobile nodes, including abundant power, processing, and storage resources; these will provide more flexibility for routing protocol design. VANET routing protocols can relax the need for energy-efficient routing mechanisms. In addition, having sufficient processing and storage resources, VANET routing protocols do not have to be compact in size and complexity.

(e) Delay Constraints

As the most common applications supported by VANETs are related to safety, VANET applications often impose hard delay constraints. Routing protocols should ensure continuous connectivity for such applications to avoid incurring any delays due to disconnections and expedite the connection setup times to keep the transmission delay as low as possible.

(f) Availability of Information Providers

The vehicle's sensor readings can be utilized in the routing protocols to enhance their functionalities. For example, GPS position information and the vehicle's speed obtained from the speedometer can be used to assist in designing efficient location-based routing protocols. Unlike the other types of networks where position and velocity information require adding special components to the nodes, most vehicles already have these components built-in. In addition, with the availability of the on-board unit that can have access to navigation software and road maps, these sources of information can help routing protocols to make better informed decisions regarding the optimum paths.

All these distinguishing challenges and features should be considered in designing a routing protocol for VANETs to be as efficient as possible and to meet the requirements and needs of the vehicular environment; the most important design considerations are handling the highly dynamic topology and intermittent connectivity to maintain connectivity among vehicles.

As shown in the discussion in this section, each network paradigm has certain design considerations that distinguish its routing requirements from the other paradigms and impose designing a distinctive set of routing protocols to meet such requirements and performance goals.

3.3 Classification and Directions

In this section, we discuss the popular classifications of the routing protocols for each network paradigm.

Mobile Ad-Hoc Networks

The most popular classification of routing protocols in MANETs is the classification according to how the route is discovered. For discovering a route or path, the routing protocol can follow the *proactive*, *reactive* or *hybrid* techniques, as shown in Fig. 3.1.

Wireless Sensor Networks

As hierarchical and position-based routing schemes are common in WSNs to enhance scalability and improve energy efficiency, it is preferable to classify routing protocols according to network structure as illustrated in Fig. 3.2. The routing protocols can be classified as *position-based* or *topology-based* protocols.

- **Topology-Based Routing**

  Topology-based routing depends on the use of information about the links and edges connecting the nodes in establishing routes. It can be further classified into *flat* and *hierarchical* routing.

  – *Flat Routing*: In flat networks, all nodes are in the same level and they all play the same role and collaborate together to perform the sensing task [2].

Fig. 3.1 Classification of
MANET routing protocols

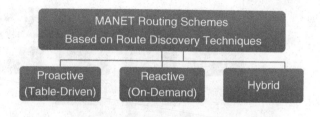

Fig. 3.2 Classification of
WSN routing protocols

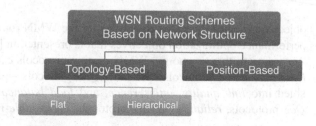

– *Hierarchical Routing*: In hierarchical protocols, nodes are divided into structured layers (such as clusters and trees), with the nodes in a layer collecting data from a lower layer, aggregating this data to reduce the number of transmissions and then sending the aggregated data to an upper layer.[2] In this type of network, nodes are usually assigned different roles depending on their layer of operation and they may have different capabilities based on their role in the network.

• **Position-Based Routing**

Position-based routing protocols depend on the use of position information for forwarding packets. In this type of network, nodes are addressed by their geographical location and it is assumed that the nodes are aware of their locations by using GPS or other localization techniques. To determine the current position of the intended destination, the source node uses a location service to get the position and includes it in the packet's destination address. Mobile nodes register their current position with the service [3].

Wireless Mesh Networks

Routing protocols in a WMN can be classified with different taxonomies. For example, protocols can be classified based on: route discovery (proactive, reactive, or hybrid), network variations (static or dynamic), protocol management (distributed, centralized, or hybrid), etc. Since the main goal of a WMN routing protocol is providing guaranteed QoS and satisfying the network performance

---

[2] It may be the sink or another layer for further aggregation.

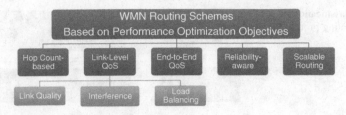

Fig. 3.3 Classification of WMN routing protocols

objectives, one of the best taxonomies for the WMN routing protocols is based on performance optimization objectives that is presented in [4] and shown in Fig. 3.3. According to this taxonomy, WMN routing protocols can be classified into either *hop-count* based protocols, *link-level QoS* protocols—which can be further classified into *link quality*, *interference*, and *load balancing* protocols-, *end-to-end QoS* protocols, *reliability-aware* protocols, or *scalable-routing* based protocols.

- **Hop-Count Based Routing**

  Protocols under this category utilize the hop-count metric for the optimum route selection. These protocols have the advantage of simplicity but they usually fail to achieve the desired QoS level.

- **Link-Level QoS Routing**

  These protocols measure quality of a route on a hop-by-hop basis at the link level. Their objective is minimizing the accumulated or bottleneck link-level effects by considering the status of each link along the path. For measuring link quality, these protocols depend on the use of parameters such as packet loss rate, retransmission count, or transmission time. This type of routing optimizes the performance objectives related to the link quality, interference, or load balancing [4].

- **End-to-End QoS Routing**

  These protocols consider the route QoS on an end-to-end basis. Such protocols achieve better end-to-end performance than the link-level ones. They depend on measuring end-to-end parameters such as delay, bandwidth, and packet delivery rate.

- **Reliability-Aware Routing**

  The objective of this category of protocols is assuring reliability of routing by applying different approaches. The most common approach is utilizing multipath routing. Multiple paths will be available to provide fault tolerance for the routing process and/or traffic distribution over the entire network.

- **Scalable Routing**

Protocols under this category are designed for large scale WMNs. Scalability of a routing protocol can be supported by following different routing approaches such as the hierarchical and geographic approaches [4].

Vehicular Ad-Hoc Networks

VANET routing protocols can be either MANET protocols that are found/amended to be suitable for the vehicular environment and characteristics, or specifically-designed protocols that are designed with the VANET challenges and features considered. Based on this view, VANET routing protocols can be classified either to be *topology-based* or *position-based* protocols. This classification is illustrated in Fig. 3.4.

The *topology-based* protocols are the MANET protocols that are suggested to be suitable for use with VANETs. As they are not designed specifically with the VANET requirements and features in mind, the topology-based routing protocols do not prove to be efficient compared to position-based protocols.

The *position-based* routing protocols prove to be the best candidates for the VANET routing functions. Some of these protocols are inherited from some MANET ones and others are newly-designed. With remarkable performance improvements over topology-based schemes, position-based routing is most commonly used and studied. Most of the protocols available in the literature and those that are currently being proposed are based on this routing scheme.

The *position-based* protocols can be further classified into *delay-tolerant* protocols and *non-delay-tolerant* protocols:

- **Delay-Tolerant Protocols**

These are the protocols designed for applications that can have some delays in delivering packets without affecting performance requirements. The packets can be kept in buffers at the nodes and delivered at a subsequent time [5, 6].

- **Non-Delay-Tolerant Protocols**

These are the protocols designed for the delay-sensitive applications. These protocols should deliver the packets in a timely fashion; otherwise, the objectives of the applications will fail [7].

Fig. 3.4 Classification of VANET routing protocols

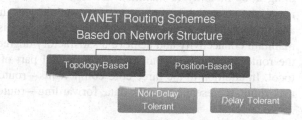

As shown in this section, each network paradigm has a unique popular classification of its routing protocols. Such a classification is derived from the unique
features of each paradigm. The basic classification is the one based on route
discovery and this is the one used for classifying the routing protocols of MANETs.
As proactive routing is not common in WSNs due to its control packet overhead,
the WSN routing protocols are commonly classified according to a different
taxonomy. The hierarchical and position-based protocols are common in WSNs
due to their scalability features; therefore, the best classification for the routing
protocols of WSNs is based on the network structure. For WMNs, as the main
objective is achieving a satisfactory QoS level and meeting desired performance
goals, the best taxonomy for the WMN routing protocols is the one based on the
performance optimization objectives. Lastly, as VANETs utilize the position-based
protocols heavily because they are the most suitable ones for their addressing
scheme—location-based—and because they also inherit some topology-based
protocols from MANETs, the best taxonomy for VANET routing protocols is the
one based on the network structure.

3.4 Core Components and Functionalities

Core components are inherited by all routing protocols for all wireless multi-hop
networks. In addition, routing protocols for each network paradigm may have
auxiliary components that are added to meet the requirements and challenges of
the network paradigm. Such components are crucial in some network paradigms to
provide the needed efficiency; therefore, these components will be also considered
core ones for that specific network paradigm.

In this section, we will discuss the core components of each wireless multi-hop
network paradigm and show which auxiliary components will be activated and
considered as core for that network paradigm. We will also explore different
functionalities deployed by the routing protocols for handling the operation of each
of these components and some representative routing protocols that adopt these
functionalities.

Mobile Ad-Hoc Networks

The most critical challenge in designing a routing protocol for MANETs is handling node mobility to maintain network connectivity. As the nodes are free to
move, the network topology incurs frequent changes in the links connecting nodes
and the neighborhood of each node; thus, links are usually prone to failures. To
maintain connectivity and achieve a satisfactory degree of reliability in MANETs,
the route maintenance component is a crucial part of any MANET routing protocol. In addition to the three core components—route discovery, route selection,
and route representation and data forwarding—route maintenance is activated

from the set of auxiliary components and is considered one of the core components of MANET routing.

(a) **Route Discovery**

Route discovery in MANETs involves different functionalities depending on the category of the protocol; proactive, reactive or hybrid. Also, under each category, there are various approaches for performing the route discovery stage.

Proactive protocols may differ in:

– *the kind of topology information exchanged among the nodes,*

Some protocols are based on exchange of distance vectors (e.g., Destination Sequence Distance Vector (DSDV) [8], Wireless Routing Protocol (WRP) [9]), and others depend on the exchange of link state packets (e.g., Fisheye State Routing (FSR) [10], Optimized Link State Routing protocol (OLSR) [11]).

– *the contents of the routing information,*

For example, the DSDV protocol shares information about the destinations, the number of hops to reach that destination, and the last sequence number seen from that destination to distinguish the stale routes from the fresh ones in order to avoid routing loops. The Distance Routing Effect Algorithm for Mobility (DREAM) [12] exploits the location and speed information of the nodes for routing packets and it depends on disseminating location information in the network by sending location updates. Being a cluster-based protocol, nodes in the Cluster-head Gateway Switch Routing (CGSR) protocol [13] periodically broadcasts a cluster member table which maps each node to its respective cluster-head.

– *which nodes the topology information is sent to,*

Some protocols send topology information only to their direct neighbors (e.g., DSDV, WRP, FSR, CGSR) and others flood the network with this information (e.g., OLSR).

– *how frequently this information is sent,*

Sending the routing information is done periodically or upon the occurrence of topology change; whichever comes first. Some protocols do not use a fixed update period. For example, the FSR protocol uses different routing update periods for the different entries in the routing table to reduce the size of the update messages. Updates for entries corresponding to the nearby nodes are sent with higher frequency than those which are for further nodes. Another example is the DREAM protocol where the frequency of the updates is a function of the node mobility.

– *how flooding of topology information is handled,*

Flooding is usually performed by disseminating packets to the whole network. This kind of communication is known as blind flooding and it incurs traffic overhead and may lead to congestion in the network. Some protocols try to reduce the flooding overhead by utilizing techniques to limit the range of transmission or to limit the number of forwarding nodes. For example, the OLSR protocol depends on the use of Multi-Point Relays (MPRs) to distribute the routing information through the network. The MPR set of each node contains the minimum number of direct neighbors that cover all the two-hop neighbors of that node. By using this technique, the number of nodes involved in disseminating the routing information is reduced; hence, the flooding overhead is mitigated. Another example is the Core-Extraction Distributed Ad Hoc Routing (CEDAR) protocol [14] where a subset of the nodes in the network are identified as the "core" nodes. These core nodes are determined using a distributed algorithm which ensures that each node has at least one adjacent core node. The link state information is propagated only through the core nodes.

– *the number of routing tables used by each protocol*

Protocols may utilize different number of tables to support their operation. For example, DSDV uses only one routing table where it stores the cost to each possible destination in the network along with the next hop to this destination and a sequence number that is assigned by the destination to specify how fresh the route is. Unlike DSDV, WRP maintains four different tables; a routing table, a distance table, a link-cost table, and a message retransmission list.

For route discovery, the *reactive* protocols employ different approaches for sending the route request (RREQ) packets through the network:

– Some protocols flood the whole network with the RREQ packets (e.g., the Dynamic Source Routing (DSR) protocol [15]). In DSR, when a node receives a RREQ, if it is the intended destination, it returns a route reply (RREP) packet carrying the whole accumulated route that it gets from the received RREQ. If it is an intermediate node and has a route for that destination stored in its cache, it concatenates the part of the route it has to the part in the received RREQ then sends the whole route in a RREP back to the source. Otherwise, the intermediate node forwards this RREQ to its neighbors after appending its address to the route list in the request. Usually, The RREP follows the reverse route back to the source. This is only possible if symmetric links are available. Otherwise, to send the RREP back to the source, the responding node initiates a route discovery process and piggybacks the RREP to the new RREQ.
– Other protocols try to reduce the flooding overhead by limiting the number of forwarding nodes or the dissemination area. For example, the Ad Hoc On-Demand Distance Vector (AODV) routing protocol [16] employs the ring-search technique to handle cases where the destination is quite near the source. In such cases, flooding RREQ packets through the whole network is wasteful. To handle that, the idea of the ring search scheme is based on searching larger areas successively. First, the RREQ is disseminated in the area around the source. If the destination is not found, the searching area is widened and so on until the destination is found. To control the area of searching, the TTL field of

the RREQ packet is assigned to 1 first then increased as the area of interest in which the source is centred is widened.

- Other protocols depend on exploiting the location and speed information to limit the RREQ flooding to a certain zone. An example is the Location Aided Routing (LAR) protocol [17] in which this zone is known as the request zone. In LAR, for route discovery, the source first estimates the zone where the destination is expected to be within. This zone is calculated based on the last known position for the destination, the last time to hear from that destination, the current time, and the speed of that destination. The flooding of RREQs is limited to the request zone which includes the expected zone and the location of the source node. Nodes out of the request zone discard the RREQ packets.
- Network congestion is one of the major problems that multi-hop networks suffer from especially if there are some intermediate nodes kept as a part of the routing path for multiple transmissions. Network congestion leads to performance degradation due to the increase in packet loss rate and the end-to-end delay. The Congestion Free Routing (CFR) protocol [18] was proposed to consider this problem, utilize the network resources efficiently, and keep the congestion level as low as possible. CFR depends on exchanging RREQs and RREPs for discovering a route, and considers only the non-congested neighbors for establishing a route. A node's congestion level is monitored by calculating the average queue length at each node. Another protocol whose primary objective is to avoid congested routes is the Node Centric Load Balancing Routing (NCLBR) protocol [19]. NCLBR works similar to AODV for discovering a route but with some minor changes to the format of the RREQ packets. As well, NCLBR divides nodes in a network into three types based on their role and connectivity: terminal, trunk, and normal nodes. In NCLBR, each node is responsible for avoiding congestion and diverting any over load to other routes in the network.

For *hybrid* route discovery, most protocols depend on deploying a proactive technique for reaching local nodes and a reactive one for reaching remote nodes:

- For example, the Zone Routing Protocol (ZRP) [20] divides the network into zones. The zone of a node comprises the nodes that are up to h hops from that node. ZRP employs two different routing approaches for the intra-zone and the inter-zone packets. Inside a routing zone, a proactive IntRA-zone Routing Protocol (IARP) is used. For communication with nodes in different zones, a reactive IntEr-zone Routing Protocol (IERP) is used. Both the IARP and the IERP perform route discovery as specified by the used proactive and reactive protocols, respectively.
- Another example is the Zone-based Hierarchical Link State (ZHLS) routing protocol [21] which assumes that the nodes know their physical location and divides them into zones based on the geographical information. Each node periodically broadcasts information about its neighbors to the nodes in the same zone and this information is stored in an intra-zone routing table. When a node has a packet to send, it checks first its intra-zone table. If the destination is out of

its zone, the source node sends a location request to all other zones via gateways. When a gateway of the zone where the destination is receives this location request, it replies to the source with a location response containing the zone ID of the destination.

(b) Route Selection

Most protocols depend on the use of only one selection metric that is determined based on the application needs. Some protocols may utilize multiple selection metrics that can be combined by means of an optimization function.

In the *proactive* protocols, route selection is done implicitly with the exchange of the up-to-date route information. When a node receives the topology information packet, either distance vector or link state, it updates the routes stored in its routing table according to the received updates.

The process of route selection in the *reactive* protocols is an explicit process that may be handled using one of three approaches: *source-based*, *destination-based*, or *intermediate-based* selection:

- An example of a protocol utilizing *source-based* route selection is the DSR protocol. In DSR, the destination replies to all RREQs it gets; then it is the responsibility of the source, after getting the RREP packets, to pick one path for forwarding the data through.
- The Load-Balanced Ad-hoc Routing (LBAR) protocol [22] utilizes *destination-based* route selection. It defines a new metric for route selection known as the degree of nodal activity which is defined as the number of active paths through a node. Also, it defines *Traffic Interference* as the sum of the neighboring activity of the current node. During its route discovery phase, each intermediate node along a potential path to the destination calculates its nodal activity and traffic interference and these values are added to the path cost obtained from preceding nodes. When the destination receives the routing information, it chooses the path with the minimum cost, i.e., least activity. Another protocol that utilizes destination-based selection is the Zone-Based Routing (ZBR) protocol [23] in which the destination picks the path with the highest stability. It depends on dividing the network area into non-overlapping square zones and determines the path stability based on the mobility factor of the nodes along the path.
- As an example of an *intermediate-based* route selection protocol, the Greedy Perimeter Stateless Routing (GPSR) protocol [24] utilizes hop-by-hop route selection. It is based on greedy routing in which the neighboring node closest to the destination is chosen to be the next forwarding node.

For the *hybrid* protocols, route selection is a combination of the route selection techniques utilized by their underlying proactive and reactive parts.

Most of the aforementioned protocols depend on only one or two parameters for route selection. Researchers realized that depending only on a few restrictive parameters does not always help in finding the ideal route. Therefore, there is an interest in developing routing protocols that consider as many affecting factors as possible with the aid of soft computing techniques. The routing protocol proposed in [25] follows this view by utilizing a neural network-based approach with the cooperation of fuzzy logic for combining a set of input parameters, and generating a set of solutions with minimal searching using a genetic algorithm. Many parameters are considered as inputs to the fuzzy logic algorithm for achieving an efficient route establishment. These parameters include: number of nodes in the system; a node's mobility across a geographical region; a node's communication capabilities; congested, blocked, and active nodes; and a node's failure history.

(c) Route Representation and Data Forwarding

Route representation and data forwarding can be handled using two different techniques: exact route and route guidance. The exact route technique can be performed using either *source routing* or *routing tables*.

- An example of the protocols that deploy the source routing approach is the DSR protocol. DSR is one of the most well-known protocols of MANETs and many other protocols inherited the idea of source routing from it. Another example is the Cluster Based Routing Protocol (CBRP) [26] which exhibits the idea of the DSR routing protocol. Unlike DSR which is a flat protocol, CBRP deploys source routing in a clustering-based fashion; only the cluster-head addresses are recorded in the accumulated address. The cluster-heads can communicate via the gateway nodes which are the nodes in the overlapping area between the clusters. The information about the adjacent clusters and the gateways to these clusters is stored in a cluster adjacency table.
- There are many protocols that depend on the use of routing tables for route representation and data forwarding. Examples are all the proactive routing protocols (e.g., DSDV, WRP, FSR, OLSR, CGSR). Some reactive protocols also depend on the use of routing tables for storing routes and next hops for the route reply and data packets and use this stored information to forward these packets. An example is the AODV protocol in which a node receiving a RREQ records the address of the node from which it received the first RREQ in its route cache to use it in returning the route reply back to the source. This is how the reverse path is created. Also, while returning the RREP back to the source, all nodes along the path keep an entry for the node from which it received the RREP and this is how the forward path is created.
- The GPSR protocol is an example of the protocols that utilize route guidance techniques. GPSR is a position-based protocol that adopts greedy packet forwarding to send a packet to a specific destination.

(d) Route Maintenance

Route maintenance is handled differently according to whether the routing protocol is proactive, reactive, or hybrid. In the **proactive** protocols, route maintenance is done implicitly with the exchange of routing updates. Having up-to-date routing information at each node helps in handling any changes in the network topology.

In the *reactive* protocols, discovered routes are usually stored in a route cache in the nodes that are involved in a transmission on those routes. These routes should be refreshed for the nodes to ensure their validity. Also, they should be invalidated when expired in order not to use them in subsequent transmissions. Also, failures of these routes should be handled either locally or by the source nodes if these routes are still needed for further transmissions. The reactive protocols use various approaches to support all the processes related to the route maintenance operation. For example:

- In DSR, each node should make sure that the packet is received by its next hop by means of acknowledgment.[3] Otherwise, the node should retransmit the packet (up to a certain limit). When the limit of retransmission is reached, the sending node should look for an alternative path to the destination in its cache. If there is no other alternative, the sending node sends a Route Error packet to the source with information about the broken link. The source should remove the broken link from its cache and start the route discovery process again.
- In AODV, to maintain local connectivity, if a node has not sent any packets to its active neighbors for a predefined period, the node broadcasts a Hello message to inform its neighbors that it is still there and alive. If any node along a route has moved, this can be detected by its predecessor that will generate a link failure notification message and sends this message to the active neighbors of this route. These neighbors send this message to their active neighbors and so on until it reaches the source which can reinitiate the route discovery process if it still needs to send data to that destination.
- The CBRP protocol deploys a Local Repair mechanism. With this mechanism, if a node detects that its next hop is unreachable, it tries to salvage the packet. It checks the source information in the packet to get the hop after the next,[4] then it checks its neighbor table and looks for a neighbor node that can reach the next hop or the hop after the next. If any of these two hops is reachable by one of its neighbors, the packet is forwarded through the new route.
- The Temporally Ordered Routing Algorithm (TORA) protocol [27] depends in its operation on creating a Directional Acyclical Graph (DAG) from the source to the destination. When a node loses its downstream link, it generates a new reference level and broadcasts the reference to its neighbors. TORA also involves a Route Erasure mechanism for retiring the invalid routes. The erase operation floods CLR packets through the network to erase invalid routes [28].

As the *hybrid* protocols are a combination of both reactive and proactive protocols, route maintenance in the hybrid protocols is also a combination of the

---

[3] Acknowledgment can be either provided by the link layer protocol or by passive acknowledgment.

[4] CBRP employs source routing, so, the whole path is stored in the data packet.

route maintenance techniques supported by the utilized proactive and reactive protocols.

Wireless Sensor Networks

The most critical design challenge for WSNs is the energy efficiency. Energy conservation techniques should be included in any WSN routing protocol to be suitable for the characteristics of the WSN, to provide efficient operation and routing functions, and to prolong the network lifetime to be operational for many years without intervention from the network designers who, in many cases, will not be able or willing to recharge the network nodes.[5] Therefore, in addition to the three core components, the energy efficiency component will be activated and considered a core one.

The WSN has its own unique data-centric addressing scheme based on the attribute-based addressing. In WSNs, the interest is on the data itself and the area where it is sensed, not on which nodes are reporting this data. Therefore, establishing routes is not based on finding a path to a specific destination with a specific address out of many. There is a sole destination which is the BS or the dedicated data collector that collects the reported data for further analysis. The location of the BS is known and, in most cases, it is unchangeable throughout the whole operation of the network. The focus is not on discovering and selecting an optimum route; the focus is on reporting and forwarding the data to the BS. Route discovery and selection is done in parallel with the data forwarding stage. Since the operation of these stages is combined together, we propose merging the three core components—route discovery, route selection, and route representation and data forwarding—into one core component which we call Route Establishment and Data Forwarding. With the introduction of this new component, the core components of the WSN routing protocol are the Route Establishment and Data Forwarding component and the Energy Efficiency component.

(a) Route Establishment and Data Forwarding

Communication in WSNs can be either (1) data-based with the nodes commencing the communication by reporting or advertising the data they have either on a time or event-driven basis, or (2) query-based with the BS issuing data interests and disseminating them through the network asking for specific sensing tasks. Based on these communication patterns, we classify the route establishment and data forwarding techniques into either *source-initiated* or *sink-initiated* techniques:

---

[5] For example, it may be infeasible or impossible to change the batteries for the nodes deployed in harsh or hostile environments.

(1) **Source-Initiated**

Many protocols depend on the source-initiated techniques and these protocols are typically designed for applications like the data gathering or object tracking applications. Some of these protocols send their actual data to the data collector without any prior negotiation and others deploy, preceding the actual data transmission, a kind of meta-data advertisement for conserving the energy that may be wasted in sending and receiving data that is not of interest to others. What follows are examples of protocols that utilize source-initiated techniques:

- *Actual Data Transmission Based*

  - The Low Energy Adaptive Clustering Hierarchy (LEACH) protocol [29] is the first and most popular energy-efficient hierarchical clustering algorithm for WSNs. Nodes in LEACH are partitioned into clusters with each cluster member reporting data periodically to its cluster-head (CH). Each CH collects the data from its cluster members, then aggregates and sends it directly to the BS. The election of CHs is a distributed, stochastic, energy-aware process that is run locally at the nodes. Heinzelman et al. also presented a variant, called LEACH-Centralized (LEACH-C), which moves the burden of the CH election to the BS.
  - The Minimum Cost Forwarding Algorithm (MCFA) [30] exploits the fact that the destination is always known; it is the BS. In MCFA, each node maintains the least cost path estimate from itself to the sink. The source broadcasts the data to its neighbors. A node receiving a packet will rebroadcast it if the node is on the least cost path between the source and the sink.
  - The Power-Efficient Gathering in Sensor Information Systems (PEGASIS) protocol [31] is a hierarchical chain-based protocol. The chain in PEGASIS is the set of nodes that are closest to one another and form a path to the BS. Each node sends its data only to its next node in the chain, the nearest one to it. Each chain has a chain leader that collects the data from other nodes in the chain, then aggregates and sends it directly to the BS.
  - To improve the performance of the traditional hierarchical protocols such as LEACH and PEGASIS, a trend of engaging artificial intelligence techniques in establishing a route and forming the network hierarchy is being adopted. Some examples of these intelligent hierarchical protocols that involve actual data transmissions are the protocols proposed in [32, 33]. In Ref. [32], Kumar et al. proposed a routing protocol based on a genetic algorithm (GA) that handles the clustering and network configurations as an optimization problem. The goal of their proposed protocol and use of a GA is to minimize the number of cluster-heads for less channel contention and higher efficiency. The protocol divides the network into a number of independent clusters using a GA that determines the number of clusters, cluster-heads, members of each cluster, and the transmission schedules for a given number of transmissions. Simple heuristics are used to keep energy efficient clusters for a longer time than less energy efficient clusters. The clustering setup and configurations are

handled in a centralized manner by the BS that broadcasts these details to all the sensor nodes. In Ref. [33], Kalantari et al. proposed a protocol that depends on the use of AI and soft computing techniques as well. It treats the network as a multi-agent system so each node is considered an intelligent agent that has a specific task. It depends on the use of machine learning techniques and a Q-learning algorithm to handle route setup. As well, a fuzzy technique is used to handle agent/node rewards and prioritization for efficient relay selection based on the nodes' residual power and distance to the destination.

- *Meta-Data Transmission Based*

  – The Sensor Protocol for Information via Negotiation (SPIN) [34] is a family of flat protocols that depends on the idea of data dissemination through the whole network assuming that all nodes are possible BSs. SPIN uses an intelligent data dissemination mechanism to reduce the receiving of redundant and unnecessary data from sensors monitoring overlapping areas and to avoid the implosion problem. It achieves this by using a data negotiation algorithm. Nodes in SPIN form meta-data that represents the actual data but with much less size and is sent in an ADV message. SPIN is a three-stage handshake protocol that includes the use of three messages that represent the three stages: ADV, REQ, and DATA. The family of SPIN protocols can also be suitable for an environment with mobile sensors as data is disseminated through the whole network.
  – Another example is a variant of the LEACH protocol called LEACH with negotiation [29] which inherits the main idea of SPIN. In LEACH with negotiation, cluster members precede the actual data transmission with a negotiation phase with its CH to ensure that only new data is transferred to the CH to avoid data redundancy.

(2) **Sink-Initiated**

Sink-initiated protocols are based on the query/interest-based communication pattern. These protocols involve different functionalities based on how queries/interests are sent. Queries can either be flooded through the whole network or sent to an area of interest based on location-aware mechanisms. In the following, we will shed light on some protocols and functionalities that utilize the sink-initiated mechanisms:

- *Flooding Based*

  – The Directed Diffusion (DD) protocol [35] is one of the most popular WSN routing protocols. It is a flat, data-centric protocol in which the interest is named as an attribute-value pair that describes a sensing task. The response is described with a similar attribute-value naming. It starts with the sink flooding exploratory interests through the whole network. When a node receives an interest, it sets up a gradient toward the neighbor from which it received the interest. So, the gradients determine the path back to the originator of the interest. Sensors that hold data matching the interest send the requested data along the gradients. In the exploratory phase, the sensed data is reported with a low rate to the sink. The exploratory phase is followed by a reinforcement phase to get the data with higher rate for more accurate event detection through specific nodes along a specific path. Choosing the neighbor that is to be reinforced can be based on many criteria. The simplest way is to reinforce the neighbor from which the first response for the interest is received. This way, the reinforced path will be the path with the lowest delay.

- The Rumor Routing protocol [36] is a variation of the DD protocol. Its key idea is to limit flooding to the nodes that have observed events rather than to the whole network. It uses agent packets and event tables to minimize the interest flooding overhead of DD.
- The Gradient-Based Routing (GBR) protocol [37] is another variant of DD. The key idea of GBR is memorizing the number of hops when the interest is diffused. Each node calculates its height which is the minimum number of hops to reach the BS. The difference between the node height and that of its neighbor is considered the gradient of the link. The link with the largest gradient is selected.
- The Reliable Energy Aware Routing (REAR) protocol [38] is based on flooding interests but instead of discovering only a single path between a source and the sink, REAR discovers an additional backup path to the same source. In discovering the backup path, only the nodes that are not involved in the service path will broadcast the backup path request to maintain two completely disjoint paths. REAR also utilizes a path reservation mechanism in which every intermediate node on the path will mark part of its energy as reserved for this path.
- The COUGAR protocol [39] separates the query processing task from the network layer and adds another layer, the query layer, between the network and the application layers. It represents the network as a huge database system with the BS being responsible for generating the query plan, defining the query flow and its in-network computation and forwarding it to the relevant nodes [2].

- *Location-Aware Based*

  - An example is the Geographical and Energy Aware Routing (GEAR) protocol [40] which is considered an improvement of the DD protocol. It exploits the fact that the interests are usually formed for a target region and they carry geographical information about it. It uses this geographical information to direct the interests to the target region without flooding. It assumes that each node has information about the location and current energy level of itself and its own direct neighbors which are exchanged by the use of periodic Hello Messages. For forwarding a packet, a node computes a cost for each of the neighbors that are closer to the target region. This cost is calculated based on the residual energy of the neighbor and its distance to the centroid of the target region. The node picks the neighbor with the smallest cost to forward the data to. Once the packet gets into the target region, it is disseminated within the region using either restricted flooding or recursive geographic forwarding.[6]

(b) Route Energy Efficiency

There are many techniques for achieving energy efficiency and reducing energy consumption in WSNs. The following are some of the common techniques utilized by protocols in the literature. One protocol may utilize multiple techniques:

(1) Use of Meta-Data

- The deployment of the meta-data approach proved to provide great savings in energy efficiency. For example, the SPIN family of protocols are able to transmit up to 60 % more data for a given amount of energy than conventional protocols [34].

(2) Data Aggregation

- In SPIN, a receiver may perform data aggregation for the received and its own data, if it has any, before advertising the data.
- The DD protocol allows data and interest aggregation. Two interests with completely overlapping attributes can be represented with a single interest entry.
- As a variant of DD, the GBR protocol also utilizes data aggregation.
- In LEACH, each CH aggregates the data received from its cluster members before sending it to the BS.
- Another example is the COUGAR protocol in which sensor nodes select a leader node to perform data aggregation and forward it to the BS.
- In PEGASIS, the chain leader aggregates the data of the other nodes in the chain and sends it to the BS.

(3) Use of Resource Manager

- One of the SPIN family of protocols, SPIN-2 (or SPIN-EC), includes a resource manager that monitors the energy consumption of the node and adapts its operation according to its available energy.
- In REAR, a resource manager at each node keeps track of the node energy and suppresses the path-request broadcasting messages to be sent by an energy-weak node.

---

[6] The recursive geographic forwarding divides the target region into four sub-regions and a copy of the packet is sent to each sub-region. The splitting and forwarding procedure is repeated till the sub-region has only one node.

(4) **Putting Nodes into Sleep Mode**

- In DD, to save power, sensors are kept off until tasked by a reception of an interest.
- Another example is the LEACH protocol which is based on a TDMA scheme. Each member sends its data to the CH in its own time slot and it can go to the sleep mode in other slots to conserve its energy.

(5) **Load Balancing**

- The LEACH protocol adapts dynamic clustering by dividing time into rounds and, at the beginning of each round, the roles of CHs are rotated to balance the energy consumption among the nodes.
- The GBR protocol deploys a traffic spreading technique to balance the traffic load among the nodes. New data streams are not passed through nodes that are currently part of the routes of other data streams.
- Another example is the Energy-Aware Routing protocol [41] which is a variant of the DD protocol. Unlike DD which enforces one path to receive data at higher rates, the Energy-Aware Routing protocol maintains multiple paths at each node, and selecting the path is based on the energy consumption level of each stored path. Having different paths selected at different times balances the load and energy consumption among the nodes in the network.
- In the PEGASIS protocol, the role of chain leader is rotated among nodes in the chain in a round robin fashion to balance the energy consumption among the nodes.

(6) **Use of Energy-Aware Metrics**

- In addition to deploying dynamic clustering, the CH election process in LEACH is based on the nodes' current residual energy.
- The GEAR protocol depends on the use of an energy-aware heuristic that considers the geographical information as well for selecting the next-hop toward the target region.

(7) **Use of Restricted Flooding**

- The GEAR protocol utilizes restricted flooding as one of the options for disseminating packets within the area of interest.

With the scope of WSNs being expanded, researchers and application developers realized that, for some applications, there is a need for getting real-time and precise information about the fast-changing events of the monitored phenomena. This need for precise monitoring drove the need for designing sensor nodes capable of capturing, storing, and sending multimedia data. Sensor networks comprised of such powerful nodes are known as Wireless Multimedia Sensor Networks (WMSNs). Generating and reporting multimedia data imposed some stringent QoS requirements that should be met by WMSNs and their protocols, including the routing protocols. WMSNs routing protocols should add to the traditional WSN protocols some techniques for handling the QoS requirements such as achieving low delay, high throughput, and a satisfactory level of reliability while keeping the main goal of improving energy efficiency. For the data forwarding techniques to be adopted by WMSNs, only the event-based and query-based techniques are considered. The continuous reporting model is not suitable for use in such networks because the continuous delivery and compression of multimedia data is very energy-consuming which will lead to a fast depletion of a node's battery. With these special routing requirements, there is an interest in developing WMSN routing protocols that consider QoS and handle the multimedia reporting challenges. Some of these protocols can be found in [42, 43]. Although QoS is considered in those protocols, it is still considered an auxiliary requirement for the traditional WSNs.

Wireless Mesh Networks

WMNs are comprised of almost stationary nodes. This feature leads to relaxing the mobility and energy constraints. Therefore, WMNs are not in need of auxiliary routing components. The WMN core components are the three main core routing components: route discovery, route selection, and route representation and data forwarding. The WMN QoS requirements and needs can be handled by special metrics and functionalities involved in the route selection component as will be discussed later in this section.

(a) Route Discovery

Some protocols, for route discovery, are proactive, others are reactive, and some are based on hybrid techniques.

(1) Proactive-based

– The Light Client Management routing Protocol (LCMP) [44] makes use of two routing tables for maintaining the topology information; a table for maintaining information about the local mesh clients and another table for recording information about the remote clients and the mesh routers associated to them.

– The Tree-Based Routing (TBR) Protocol [45] is the proactive part of the default, hybrid routing protocol proposed by IEEE 802.11s for WMNs. TBR depends on broadcasting beacon messages carrying the gateway's information to maintain a tree-like topology. TBR assumes that all traffic is either directed to/from the gateway and it maintains a multi-hop path from each mesh router to the tree root, the gateway. The drawback of TBR is that, for intra-mesh traffic, the protocol unnecessarily overloads the root which leads to scalability problems.

– The Optimized Tree-based Routing (OTR) protocol [46] improves on the TBR protocol by allowing for multiple gateways instead of the single gateway assumption of TBR. As well, OTR divides the tree into pieces and the route is calculated partly by the branches instead of wholly by the root. Having this optimized tree structure leads to less load at the root and higher support for large scale networks.

As WMNs are mainly used to cover huge areas, the number of mesh routers and clients to be maintained at each node in the network may be high. So, pure proactive routing is not preferred for use in WMNs as it will cause storage overhead and bandwidth wasting.

(2) Reactive-based

– The Link Quality Source Routing (LQSR) Protocol [47] inherits all the functionalities of the DSR protocol including the ones for route discovery (RREQs and RREPs). It only differs in how a route is selected as it supports link quality metrics. The Multi-Radio LQSR (MR-LQSR) protocol [48] works similarly to LQSR as well as supporting multi radio operations.
– Many protocols (examples mentioned later in this section) are based on the reactive hop-by-hop routing approach; no discovery for the full route should be completely done a priori. The next hop is discovered and determined on the fly based on the status of links.

(3) Hybrid

– The Orthogonal Rendezvous Routing (ORR) Protocol [49] is designed for networks that can have directional communications for their nodes. In ORR, each node can determine the position of its neighbors relative to its North. The source sends route discovery packets and the destination sends route dissemination packets in orthogonal directions and a rendezvous point is located in the intersection of these transmissions. The remainder of the discovery process is handled in a reactive way between the source and the rendezvous point and proactively for the remaining part to the destination.
– The AODV-Spanning Tree (AODV-ST) Protocol [50] inherits the AODV functionalities for the intra-mesh traffic and a spanning tree for communications to/from gateways. It is designed for the multi-radio WMNs.
– The Hybrid Wireless Mesh Protocol (HWMP) [51] is the default routing protocol of IEEE 802.11s. HWMP is a combination of an adaptation of the AODV reactive protocol with the use of radio-aware link metrics, and the TBR proactive protocol. The HWMP protocol uses MAC addresses instead of IP addresses and this is the reason the IEEE 802.11s task group preferred to use the term path selection instead of routing.

In addition, WMN routing protocols can be either centralized or distributed protocols in terms of how route discovery is handled. Most of the protocols available in the literature are distributed. A few are based on centralized techniques. For example, the Integrated Routing and MAC scheduling Algorithm (IRMA) [52] is a cross-layer, centralized, interference-based MAC-Routing protocol with a TDMA access mechanism. In IRMA, nodes and topology information are exchanged in a global control plane on a dedicated channel or dedicated time slot [4].

(b) **Route Selection**

Route selection is where WMNs are distinguished from the other wireless multi-hop networks. It is considered the most important component in WMN routing. Depending on how the route is selected, this will affect how the protocol and the network as a whole will satisfy the performance requirements and QoS guarantees.

As the WMN has its own objectives such as satisfying a reasonable level of QoS and load balancing, many unique routing selection metrics are introduced for use in WMNs. They are mostly related to link quality, interference (both intra-flow and inter-flow), reliability, and other criteria that are indicators of the path quality and suitability for providing satisfactory QoS.

In Table 3.1, we illustrate the common route selection metrics introduced and used by the WMN routing protocols available in the literature [4, 53].

The most common technique for measuring metric parameters is sending probe packets. Probe packets can be sent either in a unicast or broadcast mode. Generating and sending probe packets can be done either actively, passively, or in a cooperative approach. In "active" probing, special probe control packets are generated and exchanged for this purpose. In "passive" probing, data packets can be utilized for the probing purposes too, so no extra overhead is needed. In "cooperative" probing, a node overhears data packets transmitted by its neighbors to estimate the link quality to each neighbor. Active probing is considered the most common measurement technique [57].

(c) **Route Representation and Data Forwarding**

Some protocols are based on exact route representation and others utilize some route guidance for data forwarding.

(1) **Exact Route**

- As the LQSR protocol is a variant of the DSR protocol, it makes use of the source routing approach.
- The Multipath Mesh (MMESH) protocol [58] is based also on the source routing approach as well as allowing the source to have multiple paths for reliability purposes.

(2) **Route Guidance**

- The Extremely Opportunistic Routing (ExOR) protocol [59] follows the self-routing approach so does not have an explicit routing path before data transmission starts. In ExOR data packets are buffered and broadcast in batches. Among the nodes that receive this batch, only one node will be selected to forward it and this is what is known as the opportunistic routing.

Table 3.1 WMN routing metrics

Metric	Measures and selection	Sample protocol
Hop count	Measure number of hops from the source to the destination	LCMP
	Selection of the path with the minimum number of hops	
Expected transmission count (ETX)	The expected number of transmissions to successfully deliver a packet over a link	ExOR
	To compute ETX, each node periodically broadcasts probes containing the number of received probes from each neighbor	
	Selection of the path with the least sum of ETXs of the links along the path	
Expected transmission time (ETT)	The expected time to successfully deliver a packet over a link	AODV-ST
	ETT adjusts ETX by incorporating the throughput into its calculation	
	Selection of the path with the least sum of ETTs of the links along the path	
Weighted cumulative ETT (WCETT)	An extended version of ETT	MR-LQSR
	Limited to multi-radio mode	
	Takes into account the use of multiple channels	
	Captures the transmission time on the bottleneck channels	
	Selection of the path with the least WCETT	
Effective number of transmissions (ENT)	Built on top of ETX	Quality-aware routing [54]
	Considers the mean and variance of the packet loss ratio to project physical-layer variations	
	Selection of the path with the least ENT	
Metric of interference and channel-switching (MIC)	Considers both inter-flow and intra-flow interference	Load and interference balanced routing algorithm (LIBRA) [55]
	Calculates its value based on the ETT metric	
	Adds the number of interfering nodes to the ETT value to compute inter-flow interference	
	Computes a channel switching cost to measure intra-flow interference	
	Selection of the path with the least MIC	

(continued)

Table 3.1   (continued)

Metric	Measures and selection	Sample protocol
Bottleneck link capacity (BLC)	Indicates the residual capacity of the bottleneck link of a routing path Based on the expected busy time (EBT) of transmitting a packet over a link EBT can be measured based on the packet loss rate and transmission mechanism in the MAC layer The residual capacity of a link is defined as the ratio between the idle time and EBT Selection of the path with the largest BLC	Capacity-aware routing (CAR) [56]

- The Resilient Opportunistic Mesh Routing (ROMER) protocol [60] also follows the opportunistic routing approach. In ROMER, each packet carries a cost that is set by the source node and decremented with each hop according to the WMR cost. Duplicates of the packet may be sent through the network if the packet has enough credit at many forwarders.

Some protocols apply network coding for their routing functions to reduce the transmissions and utilize the bandwidth efficiently. In network coding, each transmission carries the information of multiple packets coded all together and decoded at the destination. It can be either applied to packets belonging to the same data flow (intra-flow network coding) or packets belonging to different data flows (inter-flow network coding) [57]. An example of the intra-flow network coding routing protocols is the MAC-independent Opportunistic Routing and Encoding (MORE) protocol [61] and an example of the inter-flow ones is the Distributed Coding-Aware Routing (DCAR) protocol [62].

Vehicular Ad-Hoc Networks

The most important challenge for designing a VANET routing protocol is maintaining connectivity while nodes—vehicles—are moving at high speeds causing frequent topology changes. For that reason, route maintenance is required to be included in all VANET routing protocols and the route maintenance component will be activated and considered a core component in VANETs. By the inclusion of the route maintenance component as a core one, the core routing components for VANET routing protocols are the route discovery, route selection, route representation and data forwarding, and route maintenance components.

Following is a discussion of the common functionalities for each component and some examples of routing protocols that utilize these functionalities. In our discussion, the focus is on the position-based functionalities and protocols as these are the dominating ones in VANETs.

(a) **Route Discovery**

Before discovering a route, the destination location (either the location of a specific node or the centroid of an area of interest) needs to be known to the source. Many location services are available now and they can be accessed easily via the OBU, such as:

- Acquiring the destination position by disseminating query messages. When a destination receives a query message asking for its position, it replies to the source with a response including its current position.
- Use of distributed location services to which the nodes periodically send updates about their position and velocity vectors. A source can consult these location servers to obtain the current position of a specific node [1].

After obtaining the intended destination position, the process of finding a route to that destination comes next. VANET routing includes different functionalities for discovering routes. We can consider VANET protocols in the literature to be all reactive ones. As there are huge numbers of nodes in the network and these nodes' connectivity is highly changing, the proactive protocols will not be feasible solutions as there will be a great overhead for recording the routing information of such large network topologies at each node and the updating process will be bandwidth consuming as it will be very frequent.

Some of these protocols depend only on the control messages that can be exchanged among the nodes for establishing a route and we call them *autonomous* protocols. Others depend on utilizing other navigation and traffic information sources for the route establishment and we call them *information-assisted* protocols. Examples of navigation/traffic information sources are street maps, traffic maps, navigation software, traffic reports from RSUs, etc.

(1) **Autonomous Protocols**

Some of these protocols depend on the exchange of periodic beacon packets (*beacon-based* protocols), while others do not (*beaconless* protocols).

- *Beacon-Based*

  - The GPSR protocol [24] is a MANET protocol suggested for use in VANETs. It depends on the exchange of 1-hop periodic beacon messages carrying information about the sending node with the most important piece of information being the node's current position.
  - Protocols depending on the exchange of traditional beacons suffer from the problem of inconsistency of the node's current position and the position announced in the beacon packet. This is because nodes keep moving and the information included in the beacon packets about their positions may be outdated. The Greedy Perimeter Stateless Routing + Advanced Greedy Forwarding (GPSR + AGF) [63] solves this problem by letting nodes include

extra information in the beacon packets about their speed, direction, and total travel time.
- The Greedy Routing with Abstract Neighbor Table (GRANT) protocol [64] depends on the idea of extended neighborhood knowledge where each node knows about its x-neighborhood. To avoid the overhead of exchanging x-hop neighbor information, GRANT divides the plane into areas and assigns only one representative for each area.
- The GpsrJ+ protocol [65] uses two-hop neighbor beaconing to provide a broader view for the nodes making decisions.
- The Connectivity-Aware Routing (CAR) protocol [66] follows the DSR approach for route discovery. It only discovers the path as a list of anchor points (nodes at junctions or road curves) and this list of anchor points is stored in the packet header.

- *Beaconless*

- The Contention-Based Forwarding (CBF) protocol [67] does not depend on exchanging periodic beacon packets among the neighboring nodes. It utilizes the concept of contention among the nodes and gives a priority of forwarding for only one node. In CBF, a node holding a packet broadcasts it to all its direct neighbors. Based on its distance to the destination, each node that receives the packet sets a timer for rebroadcasting the packet with the nearest node having the shortest timer. The actual forwarder is the nearest neighbor and the other potential forwarders are suppressed [7].

(2) Information-Assisted Protocols

As mentioned above, these protocols import street or traffic information from external sources to help in forming more efficient routes. All the surveyed information-assisted protocols are also beacon-based ones. Examples are:

- The Geographic Source Routing (GSR) protocol [68] assumes the availability of city maps for its operation. It runs the selection algorithm on the map-based graph (i.e., the set of available junctions).
- In addition to utilizing static maps, the Anchor-based Street and Traffic Aware Routing (A-STAR) protocol [69] depends on the use of real-time traffic information. A-STAR utilizes two types of maps: a statically rated map (one based on stable bus routes) and a dynamically rated map (one based on real-time traffic conditions retrieved from monitoring RSUs) [7].

(b) Route Selection

Most position-based routing protocols are based on the concept of greedy routing; a node holding a packet forwards it to the neighbor closest to the destination. Since the decision is done at the node getting the packet, route selection in the position-based greedy routing protocols is considered intermediate-based.

In making routing decisions, protocols can be either *non-overlay* or *overlay* based:

(1) Non-Overlay Based Protocols

In this category of routing, all nodes can be involved in the decision making process with equal roles and functionalities. Some of the protocols based on non-overlay routing are:

- The GPSR and GPSR + AGF are examples of protocols utilizing the classic greedy routing approach for selecting the next forwarding node (i.e., picking the neighbor closest to the destination).
- The GRANT protocol depends on the use of extended greedy routing. As each node keeps information about its x-hop neighborhood, it has a new metric for selecting the next forwarding node.

(2) Overlay Based Protocols

Overlay-based protocols depend on the use of representative nodes for the routing operation overlaid on top of the real network. These nodes have special roles and, in most cases, they are responsible for the routing decisions. In VANETs, these nodes are those at the junctions as junctions are the best places for making routing decisions as there are many options to follow there. Following are some of the functionalities utilized by the overlay-based protocols:

- The Greedy Perimeter Coordinator Routing (GPCR) protocol [70] utilizes the idea of greedy routing for forwarding packets along a road segment. When a packet reaches a junction, it stops there for deciding which road segment is best to follow. The reason behind this is preventing the packet from going in a wrong direction that will add extra unfavourable delay. It gives priorities to the nodes at the junctions—coordinators—as they have more available options and a better view.
- The GpsrJ+ protocol does not restrict packets to stop at junctions. A node holding a packet may bypass the junction if it finds, by prediction, that nodes at the junction will forward the packet along the same direction.
- Some protocols depend on applying Dijkstra's algorithm for calculating the shortest path composed of a set of junctions from a source to a destination. An example is the GSR protocol. In GSR, the algorithm can be run only once with the list of selected junction points included in the packet header or the algorithm can be rerun at each forwarding node. Another example is the A-STAR protocol that, in calculating the shortest path, also considers the traffic density of the road segments.
- The Vehicle-Assisted Data Delivery (VADD) protocol [71] is designed for delay-tolerant VANETs. It is also based on using the junction points as decision making points. At each junction, vehicles choose the outgoing road with the lowest delay. Delay can be computed using a set of linear equations based on

parameters such as road length, road density, and the average speed. After determining the next outgoing road, VADD has four variations for selecting the next forwarding node: (1) L-VADD: selects the closest node to the selected outgoing road regardless of its direction, (2) D-VADD: selects a node going toward the selected outgoing road regardless of its distance to it, (3) MD-VADD: selects multiple nodes going toward the selected outgoing road, and (4) H-VADD: combines both L-VADD and D-VADD to reduce the delay incurred in D-VADD and avoid the potential loops of L-VADD [7].

These protocols utilize the greedy routing approach for forwarder selection between the junctions with the destination is the next junction point.

To determine the overlaid nodes (nodes located at junctions), there are many approaches introduced for the *autonomous* protocols:

- The GPCR protocol depends on the use of two heuristics:

  • Whether a node has two neighbors that do not list each other as neighbors while they are in communication range of each other.
  • A correlation coefficient that relates each node to its neighbors. If the coefficient is close to 0, this means that there is little correlation among the node's neighbors, so the node is at a junction.

- In the CAR protocol, a node is considered an anchor point if its velocity vector is not parallel to the one of the previous node in the packet.

The *information-assisted* protocols can get the set of junction points extracted from the street maps or retrieved from advanced navigation software.

There are few protocols that utilize the other types of route selection techniques. For example, the CAR protocol depends on the destination-based route selection technique for determining the list of anchor points toward the destination. If the destination receives multiple RREQs, it replies to the shortest one.

(c) **Route Representation and Data Forwarding**

Most VANET position-based routing protocols are based on route guidance for data forwarding and they follow the self-routing approach; data forwarding decisions are made on the fly based on the neighbors and destination positions.

Few protocols utilize the exact route technique. As a part of the route representation of the overlay-based protocols, some protocols include the list of junction points in the packet header to follow. We can consider that a kind of source routing. So, these protocols utilize both source routing for representing the junction list and greedy routing (based on route guidance) for data forwarding between two consecutive junctions.

VANET communication can be either based on pure Vehicle-to-Vehicle (V2V) communication, or may include some assistance from RSUs in Vehicle-to-Infrastructure (V2I) communication. Based on these communication paradigms, for forwarding packets to a certain destination, VANET routing protocols can be classified into two categories based on whether they make use of the available infrastructure or not. These two categories are the *non-infrastructure-assisted* (pure V2V communication), and the *infrastructure-assisted* (a mix of V2V and V2I communications) types of protocols.

(1) Non-Infrastructure-Assisted Routing

This category is the most popular one due to the fact that the deployment of RSUs will not be widely supported in the initial phase of vehicular network deployments. Most of the protocols available in the literature ignore the assistance of RSUs and depend purely on V2V communications for relaying packets. All of the previously mentioned VANET routing protocols are examples of this category.

(2) Infrastructure-Assisted Routing

Some researchers realized the fact that getting assistance from RSUs will help in improving the efficiency of routing protocols to a high degree and they started proposing routing protocols that will make use of the limited deployment of such assisting components. Some of the researchers assumed that even if RSUs will not be widely deployed, at least a unit will be installed at each intersection for traffic management. Others assumed even more limited scenarios with few units deployed in a geographical area. Furthermore, these infrastructure-assisted protocols can be classified into *backhaul-connected* and *backhaul-isolated* protocols based on the inter-RSU connectivity.

- *Backhaul-Connected*
  - The Infrastructure-Assisted Geo-Routing scheme [72] modifies a topology-aware routing protocol by taking into consideration the connectivity among RSUs. The scheme assumes that RSUs are partially connected through the Internet; therefore, it ignores the distance among these RSUs in calculating the shortest path and represents them as a unique graph node referred to as a backbone gate. Utilizing the Internet-based backbone connectivity saves a number of hops; hence, reduces the end-to-end delay and improves reliability.
  - Following the same aforementioned approach, the Infrastructure-Assisted Routing scheme proposed in [73] utilizes RSU backhaul connectivity for data relaying with the focus of the authors on handling the buffer allocation and management challenges.

- *Backhaul-Isolated*
  - An example of this category is the Static-Node Assisted Adaptive (SADV) routing protocol [74]. As an improvement of the VADD routing protocol, the main target of SADV for utilizing RSUs and V2I communication is reducing the data delivery delay. In SADV, a packet can be buffered at an RSU available at an intersection until a forwarding vehicle is encountered on the best delivery path, while in VADD, if a forwarding vehicle is not found on the best path, a packet can be sent to a vehicle on an available, detoured path. By

obligating packet transmissions to be only through the best delivery path, SADV improves the data delivery performance.

(d) Route Maintenance

Instead of handling failures of already established routes, route maintenance in VANETs involves handling failures in establishing routes due to intermittent connectivity.

It may happen that a node does not find a neighbor that is closer to the destination than the node itself. This case is known as reaching a *local maximum*. The routing protocol should include a mechanism for handling such situations by deploying a recovery mechanism. Many recovery mechanisms are introduced for the greedy protocols:

- The GPSR protocol recovers from a local maximum by entering a perimeter mode where it follows a mechanism known as the right-hand rule. This rule states that when a node enters the recovery mode, it forwards the data to the neighbor that is sequentially counter-clockwise to the virtual line formed between the node in the recovery mode and the destination [7].
- The A-STAR technique for recovery from lack of connectivity is to recompute the path to the destination from the stuck node by temporarily marking the disconnected part of the path to be "out of service".
- The delay-tolerant protocols (e.g., VADD) recover from disconnections by having the stuck node store the packet until contact with another node. This approach is known as *store-carry-and-forward*.

The selection process returns to the greedy mode once the packet reaches a node that is closer to the destination than the node encountering the local maximum problem.

Also, because of the nodes' high mobility, the information obtained and utilized in the beginning of the transmission may change and become invalid leading to disconnectivity. Therefore, protocols should utilize mechanisms for maintaining the route and network connectivity:

- The CAR protocol assumes that route disconnection results from the destination movement.[7] It depends on the use of guard packets generated at the anchor points. When a destination changes its direction, it announces that to the nearest anchor point. When the data packet reaches the old destination's location, it will be rerouted by the guarding nodes to the new estimated position.

In Ref. [75], Paul et al. presented a comparison among the well-known VANET routing protocols by showing the pros and cons for each protocol.

---

[7] It assumes that there is no disconnection problem among the anchor points.

3.5 Summary

In this chapter, we discussed the distinguishing features of the wireless multi-hop network paradigms by presenting a detailed discussion of the design considerations and popular classification of each paradigm's routing protocols. As well, for each paradigm, we presented the core routing components, along with some functionalities of each component and some representative protocols that adopt these functionalities. We summarize the core components and functionalities to provide a clear comparison among the routing functions of the four types of wireless multi-hop network. The core components and functionalities of MANETs, WSNs, WMNs, and VANETs are illustrated in Figs. 3.5, 3.6, 3.7, and 3.8, respectively.

As depicted in the figures, there are some common functionalities for the various wireless multi-hop networks as they all share the same skeleton presented in Chap. 2. However, because each network paradigm has its own environmental features and requirements, there are some unique functionalities that distinguish each network paradigm from the other wireless multi-hop paradigms. These

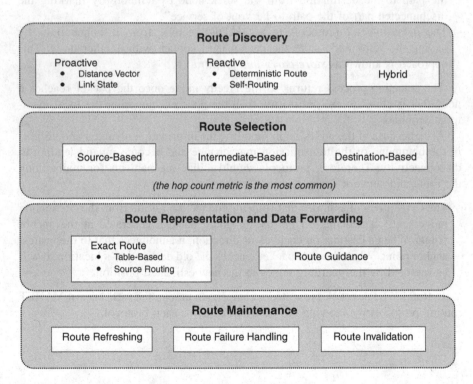

Fig. 3.5 Core components and functionalities of MANET routing

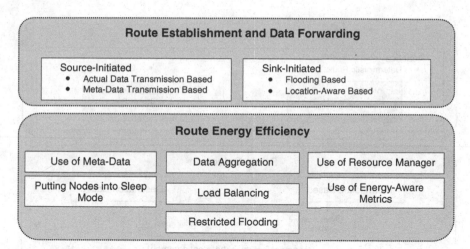

Fig. 3.6 Core components and functionalities of WSN routing

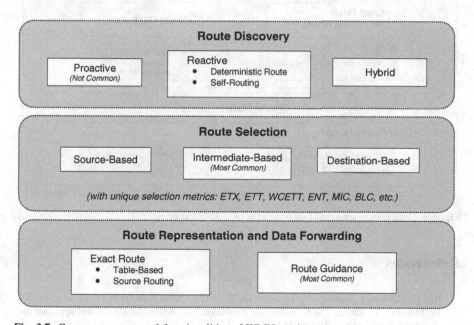

Fig. 3.7 Core components and functionalities of WMN routing

unique functionalities are added and utilized to assist the routing functions to meet the paradigm and the application performance needs.

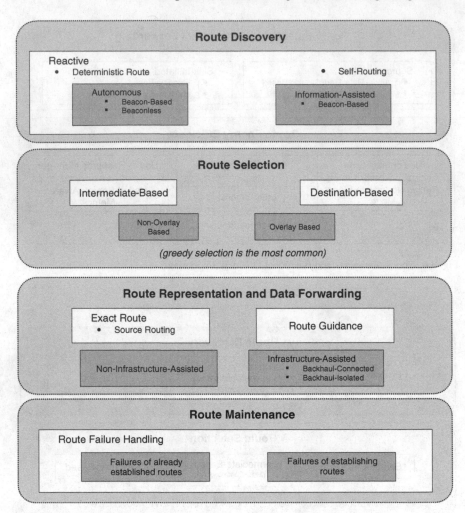

Fig. 3.8 Core components and functionalities of VANET routing

References

1. Ros FJ, Cabrera V, Sanchez JA, Martinez JA, Ruiz PM (2009) Routing in vehicular networks. In: Moustafa H, Zhang Y (eds) Vehicular networks: techniques, standards, and applications. Auerbach Publications, Boston, MA
2. Al-Karaki JN, Kamal AE (2004) Routing techniques in wireless sensor networks: a survey. IEEE Wireless Communications 11 (6):6–28. doi:0.1109/MWC.2004.1368893
3. Jin Z, Jian-Ping Y, Si-Wang Z, Ya-Ping L, Guang L (2009) A survey on position-based routing algorithms in wireless sensor networks. Algorithms 2 (1):158–182. doi:10.3390/a2010158
4. Akyildiz IF, Wang X (2009) Wireless mesh networks, vol 1. John Wiley & Sons Inc, UK

5. Wisitpongphan N, Bai F, Mudalige P, Sadekar V, Tonguz O (2007) Routing in sparse vehicular ad hoc wireless networks. IEEE Journal on Selected Areas in Communications 25 (8):1538–1556. doi:10.1109/JSAC.2007.071005

6. Huang H, Shu W, Li M, Wu MY (2008) Performance evaluation of vehicular DTN routing under realistic mobility models. In: IEEE Wireless Communications and Networking Conference, 2008. WCNC '08. pp 2206–2211. doi:10.1109/WCNC.2008.390

7. Lee KC, Lee U, Gerla M (2009) Survey of routing protocols in vehicular ad hoc networks. Advances in Vehicular Ad-Hoc Networks: Developments and Challenges, IGI Global

8. Perkins CE, Bhagwat P (1994) Highly dynamic destination-sequenced distance-vector routing (DSDV) for mobile computers. ACM SIGCOMM Computer Communication Review 24 (4):234–244. doi:10.1145/190314.190336

9. Murthy S, Garcia-Luna-Aceves JJ (1996) An efficient routing protocol for wireless networks. Mobile Networks and Applications 1 (2):183–197. doi:10.1007/BF01193336

10. Pei G, Gerla M, Chen TW (2000) Fisheye state routing: A routing scheme for ad hoc wireless networks. In: IEEE International Conference on Communications, 2000. ICC '00. pp 70–74. doi:10.1109/ICC.2000.853066

11. Clausen T, Jacquet P (2003) RFC 3626-Optimized Link State Routing Protocol (OLSR). IETF RFC3626

12. Basagni S, Chlamtac I, Syrotiuk VR, Woodward BA (1998) A distance routing effect algorithm for mobility (DREAM). In: 4th annual ACM/IEEE international conference on Mobile computing and networking, 1998. MobiCom '98 pp 76–84. doi:10.1145/288235.288254

13. Chiang CC, Wu HK, Liu W, Gerla M (1997) Routing in clustered multihop, mobile wireless networks with fading channel. In: IEEE SICON '97. pp 197–211

14. Sivakumar R, Sinha P, Bharghavan V (1999) CEDAR: a core-extraction distributed ad hoc routing algorithm. IEEE Journal on Selected Areas in Communications 17 (8):1454–1465. doi:10.1109/49.779926

15. Johnson DB, Maltz DA (1996) Dynamic source routing in ad hoc wireless networks. Mobile computing 353:153–181

16. Perkins CE, Royer EM (1999) Ad-hoc on-demand distance vector routing. In: Second IEEE Workshop on Mobile Computing Systems and Applications, 1999. WMCSA '99 pp 90–100. doi:10.1109/MCSA.1999.749281

17. Ko YB, Vaidya NH (2000) Location-Aided Routing (LAR) in mobile ad hoc networks. Wireless Networks 6 (4):307–321. doi:10.1023/A:1019106118419

18. Senthil Kumaran T, Sankaranarayanan V (2012) Congestion Free Routing in Adhoc Networks. Journal of Computer Science 8 (6):971–977. doi:10.3844/jcssp.2012.971.977

19. Ali A, Huiqiang W (2012) Node Centric Load Balancing Routing Protocol for Mobile Ad Hoc Networks. International MultiConference of Engineers and Computer Scientists 2012, IMECS '12 I

20. Haas ZJ (1997) A new routing protocol for the reconfigurable wireless networks. In: IEEE 6th International Conference on Universal Personal Communications Record, 1997. pp 562–566. doi:10.1109/ICUPC.1997.627227

21. Joa-Ng M, Lu IT (1999) A peer-to-peer zone-based two-level link state routing for mobile ad hoc networks. IEEE Journal on Selected Areas in Communications 17 (8):1415–1425. doi:10.1109/49.779923

22. Zhou A, Hassanein H (2001) Load-balanced wireless ad hoc routing. In: IEEE Canadian Conference on Electrical and Computer Engineering, 2001. CCECE '01. pp 1157–1161. doi:10.1109/CCECE.2001.933605

23. Du H, Hassanein H, Yeh C (2003) Zone-based routing protocol for high-mobility manet. In: IEEE Canadian Conference on Electrical and Computer Engineering, 2003. CCECE '03. pp 1055–1058. doi:10.1109/CCECE.2003.1226077

24. Karp B, Kung HT (2000) GPSR: Greedy perimeter stateless routing for wireless networks. In: The 6th annual international conference on Mobile computing and networking, 2000. MobiCom '00 ACM, pp 243–254. doi:10.1145/345910.345953

25. Karibasappa A, Muralidhara B (2011) Neuro fuzzy based routing protocol for mobile ad-hoc networks. In: 6th IEEE International Conference on Industrial and Information Systems, 2011. ICIIS '11. pp 216–221. doi:10.1109/ICIINFS.2011.6038069
26. Jiang M (1999) Cluster based routing protocol (CBRP). draft-ietf-manet-cbrp-spec-01 txt
27. Park VD, Corson MS (1997) A highly adaptive distributed routing algorithm for mobile wireless networks. In: 16th Annual Joint Conference of the IEEE Computer and Communications Societies, 1997. INFOCOM '97. pp 1405–1413. doi: 10.1109/INFCOM.1997.631180
28. Liu C, Kaiser J (2003) A survey of mobile ad hoc network routing protocols. Tech. rept. 2003–08, Department of Computer Structures - University of Ulm, Germany
29. Heinzelman WR, Chandrakasan A, Balakrishnan H (2000) Energy-efficient communication protocol for wireless microsensor networks. In: The 33rd IEEE Annual Hawaii International Conference on System Sciences, 2000. HICSS '00. doi:10.1109/HICSS.2000.926982
30. Ye F, Chen A, Lu S, Zhang L (2001) A scalable solution to minimum cost forwarding in large sensor networks. In: 10th IEEE International Conference on Computer Communications and Networks, 2001. pp 304–309. doi:10.1109/ICCCN.2001.956276
31. Lindsey S, Raghavendra CS (2002) PEGASIS: Power-efficient gathering in sensor information systems. In: IEEE Aerospace Conference Proceedings, 2002 pp 3-1125-1123-1130. doi:10.1109/AERO.2002.1035242
32. Kumar A (2012) Energy Efficient Clustering and Routing Techniques for Wireless Sensor Networks. International Journal of Advanced Research in Computer Science and Electronics Engineering (IJARCSEE) 1 (2):pp: 141–145
33. Kalantari S, Besheli MA, Daliri ZS, Shamshirband S, Ng LS (2011) Routing in wireless sensor network based on soft computing technique. Scientific Research and Essays 6 (21):4432–4441
34. Heinzelman WR, Kulik J, Balakrishnan H (1999) Adaptive protocols for information dissemination in wireless sensor networks. In: The 5th annual ACM/IEEE international conference on Mobile computing and networking, 1999. MobiCom '99 pp 174–185. doi:10.1145/313451.313529
35. Intanagonwiwat C, Govindan R, Estrin D (2000) Directed diffusion: a scalable and robust communication paradigm for sensor networks. In: The 6th Annual International Conference on Mobile Computing and Networking, 2000. MobiCOM '00. pp 56–67. doi:10.1145/345910.345920
36. Braginsky D, Estrin D (2002) Rumor routing algorthim for sensor networks. In: The 1st ACM international workshop on Wireless sensor networks and applications, 2002. WSNA '02. pp 22–31. doi:10.1145/570738.570742
37. Schurgers C, Srivastava MB (2001) Energy efficient routing in wireless sensor networks. In: IEEE Military Communications Conference, 2001. MILCOM '01. pp 357–361. doi:10.1109/MILCOM.2001.985819
38. Hassanein H, Luo J (2006) Reliable energy aware routing in wireless sensor networks. In: 2nd IEEE Workshop on Dependability and Security in Sensor Networks and Systems, 2006. DSSNS '06 pp 54–64. doi:10.1109/DSSNS.2006.10
39. Yao Y, Gehrke J (2002) The cougar approach to in-network query processing in sensor networks. ACM Sigmod Record 31 (3):9–18. doi:10.1145/601858.601861
40. Yu Y, Govindan R, Estrin D (2001) Geographical and energy aware routing: A recursive data dissemination protocol for wireless sensor networks. Tech. rept. ucla/csd-tr-01-0023, UCLA Computer Science Department, Los Angeles, CA
41. Shah RC, Rabaey JM (2002) Energy aware routing for low energy ad hoc sensor networks. In: IEEE Wireless Communications and Networking Conference, 2002. WCNC '02 pp 350–355. doi:10.1109/WCNC.2002.993520
42. Ehsan S, Hamdaoui B (2012) A survey on energy-efficient routing techniques with QoS assurances for wireless multimedia sensor networks. IEEE Communications Surveys & Tutorials 14 (2):265–278. doi:10.1109/SURV.2011.020211.00058

43. Zaman N, Abdullah AB (2011) Different techniques towards enhancing wireless sensor network (wsn) routing energy efficiency and quality of service (QoS). World Applied Sciences 13 (4):798–805

44. Wehbi B, Mallouli W, Cavalli A (2006) Light client management protocol for wireless mesh networks. In: 7th International Conference on Mobile Data Management, 2006. MDM '06. pp 123–123. doi:10.1109/MDM.2006.100

45. Raniwala A, Chiueh T (2005) Architecture and algorithms for an IEEE 802.11-based multi-channel wireless mesh network. In: 24th Annual Joint Conference of the IEEE Computer and Communications Societies, 2005. INFOCOM '05. pp 2223–2234. doi:10.1109/INFOCOM.2005.1498497

46. Ji Wenjiang MJ, Ma Zhuo, Tian Youliang (2012) Tree-Based Proactive Routing Protocol for Wireless Mesh Network. China Communications 9 (1):25–33

47. Draves R, Padhye J, Zill B (2004) Comparison of routing metrics for static multi-hop wireless networks. ACM SIGCOMM Computer Communication Review 34 (4):133–144. doi:10.1145/1030194.1015483

48. Draves R, Padhye J, Zill B (2004) Routing in multi-radio, multi-hop wireless mesh networks. In: The 10th annual international conference on Mobile computing and networking, 2004. MobiCom '04 pp 114–128. doi:10.1145/1023720.1023732

49. Cheng BN, Yuksel M, Kalyanaraman S (2009) Orthogonal rendezvous routing protocol for wireless mesh networks. IEEE/ACM Transactions on Networking (ToN) 17 (2):542–555. doi:10.1109/TNET.2008.926511

50. Ramachandran K, Buddhikot M, Chandranmenon G, Miller S, Belding-Royer E, Almeroth K (2005) On the design and implementation of infrastructure mesh networks. In: IEEE Workshop on Wireless Mesh Networks, 2005. WiMesh '05.

51. Bahr M (2006) Proposed routing for IEEE 802.11 s WLAN mesh networks. In: The 2nd ACM annual international workshop on Wireless internet, 2006. WICON '06. doi:10.1145/1234161.1234166

52. Wu Z, Ganu S, Raychaudhuri D (2006) IRMA: integrated routing and MAC scheduling in multi-hop wireless mesh networks. In: 2nd IEEE Workshop on Wireless Mesh Networks, 2006. WiMesh '06.

53. Campista MEM, Esposito PM, Moraes IM, Costa L, Duarte O, Passos DG, de Albuquerque CVN, Saade DCM, Rubinstein MG (2008) Routing metrics and protocols for wireless mesh networks. IEEE Network 22 (1):6–12. doi:10.1109/MNET.2008.4435897

54. Koksal CE, Balakrishnan H (2006) Quality-aware routing metrics for time-varying wireless mesh networks. IEEE Journal on Selected Areas in Communications 24 (11):1984–1994. doi:10.1109/JSAC.2006.881637

55. Yang Y, Wang J, Kravets R (2005) Interference-aware load balancing for multihop wireless networks. Tech. rept. UIUCDCS-R-2005-2526, University of Illinois, Urbana Champaign

56. Liu T, Liao W (2006) Capacity-aware routing in multi-channel multi-rate wireless mesh networks. In: IEEE International Conference on Communications, 2006. ICC '06. pp 1971–1976. doi:10.1109/ICC.2006.255059

57. Núñez-Martínez J, Mangues-Bafalluy J (2010) A survey on routing protocols that really exploit wireless Mesh network features. Journal of Communications 5 (3):211–231. doi:10.4304/jcm.5.3.211-231

58. Nandiraju NS, Nandiraju DS, Agrawal DP (2006) Multipath routing in wireless mesh networks. In: IEEE International Conference on Mobile Adhoc and Sensor Systems, 2006. MASS '06. pp 741–746. doi:10.1109/MOBHOC.2006.278644

59. Biswas S, Morris R (2005) ExOR: opportunistic multi-hop routing for wireless networks. ACM SIGCOMM Computer Communication Review 35 (4):133–144. doi:10.1145/1090191.1080108

60. Yuan Y, Yang H, Wong SHY, Lu S, Arbaugh W (2005) ROMER: resilient opportunistic mesh routing for wireless mesh networks. In: The 1st IEEE Workshop on Wireless Mesh Networks, 2005. WiMesh '05.

61. Chachulski S, Jennings M, Katti S, Katabi D (2007) Trading structure for randomness in wireless opportunistic routing. ACM SIGCOMM Computer Communication Review 37 (4). doi:10.1145/1282427.1282400
62. Le J, Lui JCS, Chiu DM (2010) DCAR: distributed coding-aware routing in wireless networks. IEEE Transactions on Mobile Computing 9 (4):596–608. doi:10.1109/TMC.2009.160
63. Naumov V, Baumann R, Gross T (2006) An evaluation of inter-vehicle ad hoc networks based on realistic vehicular traces. In: The 7th ACM international symposium on Mobile ad hoc networking and computing, 2006. MobiHoc '06 pp 108–119. doi:10.1145/1132905.1132918
64. Schnaufer S, Effelsberg W (2008) Position-based unicast routing for city scenarios. In: International Symposium on a World of Wireless, Mobile and Multimedia Networks, 2008. WoWMoM '08. pp 1–8. doi:10.1109/WOWMOM.2008.4594851
65. Lee KC, Härri J, Lee U, Gerla M (2007) Enhanced perimeter routing for geographic forwarding protocols in urban vehicular scenarios. In: IEEE Globecom Workshops, 2007 pp 1–10. doi:10.1109/GLOCOMW.2007.4437832
66. Naumov V, Gross TR (2007) Connectivity-aware routing (CAR) in vehicular ad-hoc networks. In: 26th IEEE International Conference on Computer Communications, 2007. INFOCOM '07. pp 1919–1927. doi:10.1109/INFCOM.2007.223
67. Füßler H, Hartenstein H, Widmer J, Mauve M, Effelsberg W (2004) Contention-based forwarding for street scenarios. In: 1st International Workshop in Intelligent Transportation, 2004. WIT '04. pp 155–159
68. Lochert C, Hartenstein H, Tian J, Fussler H, Hermann D, Mauve M (2003) A routing strategy for vehicular ad hoc networks in city environments. In: IEEE Intelligent Vehicles Symposium, 2003. pp 156–161. doi:10.1109/IVS.2003.1212901
69. Seet BC, Liu G, Lee BS, Foh CH, Wong KJ, Lee KK (2004) A-STAR: A mobile ad hoc routing strategy for metropolis vehicular communications. Lecture Notes in Computer Science: NETWORKING 2004, Networking Technologies, Services, and Protocols; Performance of Computer and Communication Networks; Mobile and Wireless Communications:989–999. doi:10.1007/b97826
70. Lochert C, Mauve M, Füßler H, Hartenstein H (2005) Geographic routing in city scenarios. ACM SIGMOBILE Mobile Computing and Communications Review 9 (1):69–72. doi:10.1145/1055959.1055970
71. Zhao J, Cao G (2006) VADD: Vehicle-assisted data delivery in vehicular ad hoc networks. In: 25th IEEE International Conference on Computer Communications, 2006. INFOCOM '06. pp 1–12. doi:10.1109/INFOCOM.2006.298
72. Borsetti D, Gozalvez J (2010) Infrastructure-assisted geo-routing for cooperative vehicular networks. In: IEEE Vehicular Networking Conference, 2010. VNC '10 pp 255–262. doi:10.1109/VNC.2010.5698271
73. Wu Y, Zhu Y, Li B (2012) Infrastructure-assisted routing in vehicular networks. In: The 31st Annual IEEE International Conference on Computer Communications, 2012. INFOCOM '12 pp 1485–1493. doi:10.1109/INFCOM.2012.6195515
74. Ding Y, Xiao L (2010) SADV: Static-node-assisted adaptive data dissemination in vehicular networks. IEEE Transactions on Vehicular Technology 59 (5):2445–2455. doi:10.1109/TVT.2010.2045234
75. Bijan P, Md I, Md ANB (2011) VANET Routing Protocols: Pros and Cons. International Journal of Computer Applications 20 (3):28–34. doi:10.5120/2413-3224

Chapter 4
Conclusions and Open Issues

Throughout this brief, we explored the various aspects related to routing in the wireless multi-hop network paradigms: MANETs, WSNs, WMNs, and VANETs. We presented an overview on wireless multi-hop networks along with an introduction to the four wireless multi-hop network paradigms. As well, we presented an introduction to routing, its basic functions, and how it fits in the protocol stack. We also explored the unifying features of the aforementioned networks by discussing the basic routing components that are main parts of any wireless multi-hop routing protocol and proposed a generic routing model that can be inherited in designing a wireless multi-hop routing protocol.

In addition, we explored the various features that distinguish each network paradigm from the others. We discussed the various challenges and characteristics, the most well-known routing classification, and the core routing components of each network paradigm for the sake of providing an extensive comparison between routing techniques in the four wireless multi-hop networks.

Based on our surveys and studies of the various routing functionalities and protocols of wireless multi-hop networks, we have reached a conclusion about the ideal routing protocol that provides the optimum operation with efficient utilization of the network resources. We have concluded that the self-routing approach will be the dominant one for all the network paradigms since self-routing has a great advantage in adapting to changes in the links' status and connectivity.

In MANETs, instead of keeping a deterministic route, doing routing on the fly (i.e., hop-by-hop) helps in handling links failures and disconnectivity, hence, saving the time and the bandwidth wasted in the route maintenance mechanisms and discovery of alternative routes.

In WSNs, self-routing helps in handling node failures resulting from either energy depletion or harsh environmental damage. By selecting the path on the fly, it is ensured that there will be a connected path to the destination. In addition, hop-by-hop routing helps in balancing the load among the various potential forwarders; hence, energy consumption will be well-balanced among the nodes.

S. Abdel Hamid et al., *Routing for Wireless Multi-Hop Networks*,
SpringerBriefs in Computer Science, DOI: 10.1007/978-1-4614-6357-3_4,
© The Author(s) 2013

In WMNs, it is observed that the best routing technique is the one done hop-by-hop as this technique adapts to the link status changes. Wireless links are subject to changes in their bandwidth, interference level, etc. The effects of such changes can be mitigated if the route establishment is done on the fly instead of having it deterministic. So, self-routing helps in tackling the short-term path quality variations problem of WMNs.

In VANETs, the most popular routing protocols are the self-routing ones as they are the best handlers of the intermittent connectivity and dynamic topology.

Derived from the summarizing figures presented in the previous chapter, Fig. 4.1 illustrates the ideal model and its ideal functionality per each component.

Self-routing requires that route discovery is handled in a *reactive* way, route selection should be *intermediate-based*, and the route representation and data forwarding is done based on *route guidance*.

For the selection metric, each network paradigm should have its own metric based on the goals and needs of each paradigm. MANETs can have the *mobility level* as its selection metric. Choosing the most stable neighbor–the slowest one–to be the next forwarding node reduces the probability of route failures due to links changes. For WSNs, the selection metric can be the *residual energy* of the potential forwarders. Choosing the neighbor with the highest residual energy reduces the chance of the node's battery depletion. In WMNs, there are different selection metrics that can be utilized and are suitable for the intermediate-based selection. These metrics are discussed in Table 3.1 and we can refer to them here as QoS level. Finally, the most popular self-routing based VANET protocols are based on greedy selection; in other words, the selection metric for VANETs should be the *distance to the destination*.

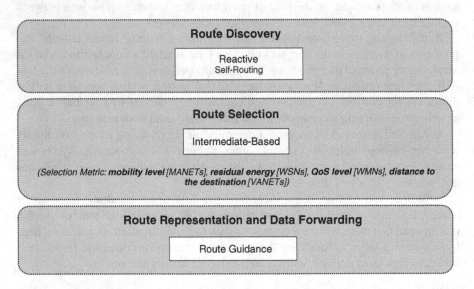

Fig. 4.1 Ideal routing model

So, by combining the functionalities shown in Fig. 4.1 and choosing a suitable selection metric, an ideal wireless multi-hop routing protocol can be designed.

As we have seen throughout this brief, a considerable number of routing protocols have been proposed for the various wireless multi-hop networks. With these protocols available in the literature, there are still many open issues that need to be covered in the area of wireless multi-hop routing. Most of the open issues are related to WMNs and VANETs considering that these are the most recent wireless multi-hop network paradigms; however, a few open issues are still there for routing in MANETs and WSNs.

MANETs are the oldest multi-hop networks and many routing protocols were proposed for use in MANETs in the past decade. Those protocols presented a variety of functionalities handling most of the challenges of MANET operation. However, some researchers are currently interested in applying some optimization and intelligent techniques in MANET routing and testing the potential improvements. Examples of such techniques are those utilizing soft computing and artificial intelligence. Some of these intelligent protocols were presented in Chap. 3. This trend for designing intelligent, optimized protocols is of interest in the design of routing protocols for all wireless multi-hop network paradigms; not only for MANETs.

The WSN paradigm is considered relatively old compared to the WMN and VANET ones with many routing protocols proposed in the literature to cover its design challenges. However, as mentioned in Chap. 3, the paradigm of wireless multimedia sensor networks is currently gaining interest from researchers. Few protocols have been proposed to serve the needs of this paradigm and it is still considered an open issue for researchers. Such protocols should consider QoS requirements while keeping energy efficiency as the main goal.

Although there are tens of WMN routing protocols proposed in the literature, there are still some open issues that need to be covered to improve the efficiency of the WMN routing function: (1) Extensive testing for the various routing metrics is needed to determine the most accurate and efficient metric or combination of metrics, (2) Proposing passive techniques for measuring metric parameters instead of depending on exchanging control packets for the sake of reducing the overhead and utilizing network resources efficiently, (3) Studying the impact of the active measurement techniques on increasing the self-interference problem, (4) Many protocols depend on the use of Dijkstra's algorithm for calculating the best path. The use of such an algorithm hinders the network scalability. Other approaches should be investigated for the optimal path computation, (5) Further exploration on the impact and the benefits of network coding strategies. Currently, only a few WMN routing protocols utilize these coding opportunities.

For VANETs, although there are many new protocols introduced for routing functionalities, routing is still considered one of the hottest topics in the VANET research area. There are some open issues and opportunities that can be handled and considered by the researchers to improve the performance of message routing and pave the road to many new applications and services to be supported by VANET communications, among them. (1) Most VANET routing protocols depend on the greedy selection which just considers the distance to the destination. This approach for routing selection lacks the consideration of some real physical conditions such as fading and interference which may affect reception of the packets by the selected recipient. Other selection metrics should be taken into consideration to provide more efficient routing decisions. Such metrics can be merged with the greedy approach by optimization solutions, (2) As VANET applications will include real time communications, for example, sharing videos and playing games on the move among passengers of many neighboring vehicles, QoS support should be considered by new VANET routing protocols to satisfy the requirements of these types of communications, (3) With the advances of the built-in VANET communication modules, many interfaces and technologies will be available for transmitting the packets. The availability of these multiple radios and technologies should be considered by the routing protocols to provide the most efficient transmission over the highly available communication technologies, (4) Routing should be aware of the higher-layer requirements and generated traffic to better assist the application operation. This leads to what is known as content-based routing in which the contents of the messages should be taken into consideration for providing the optimal routing functionalities and decisions. For example, in the case of an accident or health emergency, packets should be routed to the nearest ambulance in addition to the neighboring vehicles, (5) One of the most enabling factors in VANET communications is the inclusion of the infrastructure. Most of the VANET routing protocols ignore infrastructure exploitation for the message transmissions. Depending on the infrastructure, it can help in handling the intermittent connectivity and expediting the transmission times, hence, reducing the delay. So routing protocols that exploit the existence of the infrastructure need to be proposed. A few infrastructure-based protocols were proposed in the literature and presented in Chap. 3 but more proposals are needed to make use of the benefits gained from considering the infrastructure as a relay, (6) Network coding can be considered as an efficient mechanism for reducing the number of transmissions and utilizing network resources efficiently, (7) The number of delay-tolerant routing protocols is limited compared to the number of non-delay-tolerant ones. So further research is needed in the area of routing for delay-tolerant VANETs, (8) As VANETs can operate in many scenarios and traffic conditions and support different applications, VANET routing protocols should be adaptable to these different application requirements and be scalable to various network sizes at different times of the day.

In addition, as a common opportunity for all wireless multi-hop networks, the introduction of cognitive radios and their integrations with wireless multi-hop networks will open many research issues for wireless multi-hop routing. Cognitive radios add some challenges with their intermittent connectivity and bandwidth availability that should be handled and taken into consideration by the routing functions.

With the current focus on the integration of heterogeneous networks, a network may consist of multiple different wireless multi-hop network paradigms integrated together for cooperatively providing a designated service. Examples of such integrations are the cooperation between an on-road WSN and a VANET for better traffic management or enhanced environmental monitoring. Designing routing protocols that handle the combined challenges of many wireless multi-hop network paradigms and are able to alter their operation based on the type of node holding a packet is a hot open issue.

In short, many wireless multi-hop routing protocols are available in the literature and, although they have some common and unifying features, they also have their own distinguishing ones based on the network paradigm they are proposed for. Some hot open issues and opportunities are also available for the interested researchers to work on.